Math Mutation Classics

Exploring Interesting, Fun and Weird Corners of Mathematics

Erik Seligman

Apress®

Math Mutation Classics: Exploring Interesting, Fun and Weird Corners of Mathematics

Erik Seligman
Hillsboro, Oregon, USA

ISBN-13 (pbk): 978-1-4842-1891-4   ISBN-13 (electronic): 978-1-4842-1892-1
DOI 10.1007/978-1-4842-1892-1

Library of Congress Control Number: 2016937749

Managing Director: Welmoed Spahr
Lead Editor: Jeffrey Pepper
Technical Reviewer: Barnaby Sheppard
Editorial Board: Steve Anglin, Pramila Balan, Louise Corrigan, Jonathan Gennick,
    Robert Hutchinson, Celestin Suresh John, Michelle Lowman, James Markham,
    Susan McDermott, Matthew Moodie, Jeffrey Pepper, Douglas Pundick,
    Ben Renow-Clarke, Gwenan Spearing
Coordinating Editor: Mark Powers
Compositor: SPi Global
Indexer: SPi Global
Artist: SPi Global

Distributed to the book trade worldwide by Springer Science+Business Media New York, 233 Spring Street, 6th Floor, New York, NY 10013. Phone 1-800-SPRINGER, fax (201) 348-4505, e-mail orders-ny@springer-sbm.com, or visit www.springeronline.com. Apress Media, LLC is a California LLC and the sole member (owner) is Springer Science + Business Media Finance Inc (SSBM Finance Inc). SSBM Finance Inc is a **Delaware** corporation.

For information on translations, please e-mail rights@apress.com, or visit www.apress.com.

Apress and friends of ED books may be purchased in bulk for academic, corporate, or promotional use. eBook versions and licenses are also available for most titles. For more information, reference our Special Bulk Sales–eBook Licensing web page at www.apress.com/bulk-sales.

Any source code or other supplementary materials referenced by the author in this text are available to readers at www.apress.com/9781484218914. For detailed information about how to locate your book's source code, go to www.apress.com/source-code/. Readers can also access source code at SpringerLink in the Supplementary Material section for each chapter.

Printed on acid-free paper

Contents at a Glance

Contents

About the Author

Erik Seligman is a Formal Verification Architect at Intel Corporation, where he has been an engineer for over two decades. Erik has hosted the *Math Mutation* podcast since 2007. He was recently lead author of the well-received technical book, *Formal Verification: An Essential Toolkit for Modern VLSI Design* (Morgan Kaufmann, 2015). He earned a M.S. in computer science from Carnegie Mellon and a B.A. in mathematics at Princeton. He also has served since 2013 as an elected director on the board of the Hillsboro School District, the 4th largest K-12 school district in Oregon.

About the Technical Reviewer

Barnaby Sheppard is a mathematics writer and editor based in London, UK. Formerly a lecturer and researcher in pure mathematics (University College Dublin, Durham University, Lancaster University), he continues to pursue the subject, now branching out from the functional analytic and operator algebraic roots of his PhD and subsequent early research. He is the author of The Logic of Infinity (Cambridge University Press), an introduction to axiomatic set theory and related mathematical/philosophical matters, and is currently working on a number of other long-term writing projects. He is also interested in certain theoretical aspects of music and is a keen guitarist. He can be contacted via his book's website: https://thelogicofinfinity.wordpress.com.

Acknowledgments

I would like to thank my wife Ann and daughter Sonia, whose patience and support made this book possible.

I would also like to thank my friends and colleagues who reviewed early drafts and sent feed-back and suggestions: Timothy Y. Chow, Joe Leslie-Hurd, Winston Ou, and Bhavinkumar Patel. Thanks guys!

Playing with Mathematics

My great-aunt Evelyn used to love to relate an anecdote about babysitting me and my two younger sisters, from when I was only five years old. After chasing my sister around the living room for a few minutes, she realized she had lost track of me, and began searching the house as she called out my name. Finally she heard me answer from one of the bedrooms, "I'm in here, playing". Worried about what a 5-year-old might be finding to play with in her bedroom, she called out "Playing with what?" as she rushed down the hallway. I answered back "Playing with my mathematics!" She was relieved to arrive and find me seated quietly in the middle of the floor, scribbling numbers on a note pad.

Most listeners laugh at this story, finding the concept of "playing" with mathematics somehow absurd. But why is that? I think it's a sad failure of our modern education systems that so many of my fellow citizens think of math as a burden, something to be endured in school, something necessary for certain jobs, or at best a useful tool for engineering and technology. When I hear the word "mathematics," here are some of the things I think of:

- Strange insights you get from basic counting: did you know that nearly everyone in Pittsburgh has an above-average number of fingers?

- Bizarre quantities like π or infinity, that make less sense the more you think about them: can one infinity actually be larger than another?

- Mysterious journeys into extra-dimensional spaces: do we live in a three, four, eleven, or infinite-dimensional world?

- Mental paradoxes that result from attempts at precise reasoning: can you dispute a philosophers' proof that there is a hippopotamus in your basement right now?

- Insight into the working of our own minds: can we explain why time passes faster as we get older?

- The fundamental nature of reality: are we all really a computer simulation?

When you connect mathematics with topics like these, it suddenly grows into a much more playful and motivating topic than the rote techniques that were drilled into you in school. Ideas like these are what motivated me, almost a decade ago, to create the Math Mutation podcast. Despite the fact that there were thousands of podcasts out there (and are even more today), I found that there wasn't really any that occupied this niche: brief and focused introductions to fun, interesting, and weird corners of mathematics that you are unlikely to have heard in school. There were a few other math podcasts out there, but the primary focus tended towards homework help for current students. This is a laudable goal, but misses so much beyond that. My key guiding principles in assembling topics for the Math Mutation podcast have always been:

Accessibility. The average listener with no math education beyond high school, or even a current high school student, should be able to comprehend and enjoy the podcast.

Brevity. The idea is to give a quick tidbit that whets the appetite, sharing a single intriguing or surprising mathematical concept, and communicating the basic idea. For fully elaborated details or rigorous proofs that would satisfy a math professor, listeners can follow up and check out the meatier references in the show notes.

Welcoming. One flaw I've seen in many books and podcasts of the "interesting math" genre is the focus on presenting problems to be solved. While problem-solving is a critical tool for mathematical growth, it can be intimidating to those who are less experienced or less confident in their own abilities. Thus, I intentionally avoid this path, choosing direct presentation over problem-solving challenges.

Fun! The topic needs to be something I find unusual, fascinating, or surprising, and is typically not learned in a high school or college mathematics class.

I should also mention that in creating this podcast, I have "stood on the shoulders of giants". I grew up reading mind-bending books and essays by brilliant authors like Martin Gardner, Douglas Hofstadter, Rudy Rucker, and Isaac Asimov, and many of the topics were inspired by those writings. In general, I'm not claiming to have derived new mathematical results, but I have cherry-picked a set of especially intriguing mathematical tidbits, and provided a short and focused exposition that will whet your appetite to learn more. Hopefully this will be a launching point, leading to many further hours of enjoyment as you follow up on your favorite topics using the "References and Further Reading" chapter near the end of this book.

When assembling the podcast episodes to transfer into this printed format, I have attempted to group them into logical chapters, and have done my best to revise and improve them. I think you, the reader, deserve a little bonus content for buying this book, rather than just listening to the free episodes online. I hope that after reading this book, you too will think of mathematics as not just a required school topic, necessary burden, or useful engineering tool, but as something fun to play with.

CHAPTER 1

■ ■ ■

Simple Surprises

Many of the amusing mathematical tidbits that I've highlighted in the podcast do not require very deep or complex reasoning: they are simple consequences of logically thinking through our basic notions about numbers, counting, or probability. In this first chapter we discuss some of these very basic but somehow surprising ideas.

City of Mutants
From Math Mutation podcast 1

Did you know that almost everybody in Pittsburgh has an above-average number of fingers? Some say it's nuclear waste. Some say it's toxic pollution. But I say it's just the math.

Suppose a city has 2 million residents. Assuming everyone is healthy, we would expect them to have 20 million fingers among them. So the average number of fingers per person is 20 million divided by 2 million, or 10. But in real life, is there ever a city where everyone is healthy?

In any population, a tiny number of people will have lost a finger due to an industrial accident or over-zealous *World of Warcraft* keyboard-pounding. So suppose 10 people in the city have lost a finger. Then the total number of fingers is not 20 million, but 19,999,990. (There may be a few with extra fingers to balance this out, but that's a much rarer condition, probably negligible for the purpose of this calculation.) This brings the average per person down to 9.999995. Yet this doesn't change the fact that nearly everybody has 10 fingers, which beats the average by 0.000005! So nearly everyone does have an above-average number of fingers.

As you have probably noticed, we're playing with the difference between our casual conversational notion of "average", and its mathematical definition. The average is usually formally defined to be the *mean,* the sum of a measure divided by the size of the population, which we calculated above. But in casual conversation, we think of the word "average" as denoting a typical member of the population. In terms of formal definitions, this more closely matches the *median,* where you sort the population by some measurement, and take the middle member as a representative. The median resident of Pittsburgh does indeed have the normal 5 fingers per hand you would expect.

© Erik Seligman 2016
E. Seligman, *Math Mutation Classics*, DOI 10.1007/978-1-4842-1892-1_1

In any case, the statement that opened this episode will work for just about any city. Having lived there for four years back in the 1990s, I think Pittsburgh may indeed be a uniquely mutated city, but if so it's for reasons beyond this discussion. So if things get boring at your next social gathering, be sure to liven it up by pointing out that, using the average number of fingers as the measurement, your city is also a city of mutants.

Two Plus Two Equals Five
From Math Mutation podcast 210

Recently I heard someone quote a clever metaphor in a casual conversation, "Life is when nature takes 2 and 2 to make 5." It's a nice statement of how living creatures are more than the sum of their parts. If you took all the chemical compounds in my body and dumped them on the ground in the right proportions, all you would get is a mess. Yet somehow I am here, and at least sentient enough to record math podcasts. I went online to try to find the source of this quotation, and was surprised to see the number of references to this seemingly silly nonsense equation, $2 + 2 = 5$.

Most of us are probably familiar with the equation from George Orwell's classic novel *1984*. As you probably recall, in the book, people are told that if the government says that $2 + 2 = 5$, it is the duty of all citizens to believe it – not just say it, but actually come to believe that it is true. Surprisingly, Orwell did not come up with this out of thin air: a real-life totalitarian government, the Soviet Union, actually did use $2 + 2 = 5$ as part of its propaganda, in a poster with the title "$2 + 2 = 5$: Arithmetic of a counter-plan plus the enthusiasm of the workers." It wasn't quite as blatantly absurd as in *1984*, but the Soviet propaganda poster used it as a metaphor: supposedly a 5-year plan could be completed in 4 years, because the enthusiasm of the workers provided an invisible additive factor. Sadly, most of this "enthusiasm" was mainly due to fear of being sent to the Gulag prison camps, which resulted in many managers doctoring the statistics to match the results that the government wanted – on paper only. It's also reported that Nazi Hermann Goering actually used this metaphor in real life, once saying "If the Führer wants it, two and two makes five!"

The phrase '$2 + 2 = 5$' has actually existed in the arts from the early 19th century. According to *Wikipedia*, the phrase was first coined in a letter from Lord Byron, where he wrote "I know that two and two make four—& should be glad to prove it too if I could— though I must say if by any sort of process I could convert 2 & 2 into five it would give me much greater pleasure." He may have been making an indirect reference to Rene Descartes' *Meditations*, where the famous philosopher discussed whether equations such as $2 + 3 = 5$ exist outside the human mind, and whether they can be doubted: "And further, as I sometimes think that others are in error respecting matters of which they believe themselves to possess a perfect knowledge, how do I know that I am not also deceived each time I add together two and three, or number the sides of a square, or form some judgment still more simple, if more simple indeed can be imagined?"

Later, Victor Hugo used this concept in a critique of the mob rule that had led to Napoleon's rise to power, foreshadowing Orwell's later political metaphor: "Now, get seven million five hundred thousand votes to declare that two and two make five, that the straight line is the longest road, that the whole is less than its part; get it declared by eight millions, by ten millions, by a hundred millions of votes, you will not have advanced a step." Russian authors Ivan Turgenev, Leo Tolstoy, and Fyodor Dostoyevsky also made use of this metaphor. Turgenev used it to symbolize divine intervention: "Whatever a man prays for, he prays for a miracle. Every prayer reduces itself to this: Great God, grant that twice two be not four." In the 20th century, there were many instances of authors following Orwell's lead and again using this metaphor for the struggle against totalitarianism, including Albert Camus and Ayn Rand.

An intriguing question is whether there are cases when it is actually valid to say that $2+2=5$. A well-known mathematicians' joke is that "$2+2=5$, for particularly large values of 2." This may refer to issues with rounding: if you start, for example, with the obviously correct equation "$2.4+2.4=4.8$", and ask someone to round all the numbers to the nearest integer, you do indeed derive "$2+2=5$" from this true equation. It also might be a case of playing with the definitions of symbols: perhaps you can define the symbol that we normally write as "2" to actually be an algebraic variable representing the value 2.5. You can also find various tricky "proofs" that $2+2=5$ circulating the web, where many lines of complex algebra are often used. These many lines usually misdirect you from one invalid step, where a term t is replaced with the square root of t^2 (it should really be the absolute value of that quantity), or both sides are divided by a term that equals 0. Here is an example of one of the simpler ones:

$$4-4=10-10 : \textit{Start out with true statement}$$

$$(2-2) \times (2+2) = 2 \times 5 - 2 \times 5 : \textit{Rewrite both sides in a complex form}$$

$$(2-2) \times (2+2) = (2-2) \times 5 : \textit{Regroup factors on the right-hand side}$$

$$==> 2+2=5 : \textit{Divide both sides by } (2-2)$$

As you have probably noticed, the last step divided both sides by 0, which is not algebraically valid, and results in nonsense.

An amusing spoof article online [Eul04] points out some real-life situations where 2 and 2 might really make 5. Ancient Incas used knotted ropes to track business transactions, and if you tie together two ropes that each have two knots, the resulting rope will have 5 knots, including the one used to tie them together. Another example is if you put 2 male and 2 female rabbits in a cage – pretty soon you will see numbers way larger than 5. I'm sure that most people who experience these situations in real life can make the distinction between the messiness of reality and the related arithmetic though.

But that last example brings us back around to the quote that started this whole thing. Ironically, my web searching did not succeed in uncovering the source of the clever comparison between life and making two plus two equal five. Most likely I didn't remember the phrasing exactly right, or else someone was just coining this on the fly and it didn't really come from a famous quote. If you have heard it before and know its origin, please send me an email!

Stupid Number Tricks
From Math Mutation podcast 23

One act you could occasionally see at carnival sideshows in the 19th and 20th centuries was the 'calculating prodigy'. This would be someone who could perform amazing feats of calculation, such as instantly multiplying large numbers or finding 23rd-roots, in front of an audience. Some of these prodigies were true geniuses or autistic savants. But today we are going to talk about the fakes, people who were not geniuses, but still managed to impress audiences with their mathematical feats. Martin Gardner described many of their tricks in his classic "Mathematical Games" series in *Scientific American*.

Some of their methods were simply what we would call "cheating". The easiest way to cheat in this kind of performance is to have a confederate in the audience, who you happen to point at to call out the numbers. Then it's a simple matter to pre-calculate some difficult problem, and have the answer ready in an instant. A more legitimate cheating method is stalling for time. For example, after an audience member calls out two large numbers to multiply, ask him to repeat them to be sure you got them right, then slowly write the numbers on a blackboard. Your assistant will yell 'Go' and start the stopwatch after you have finally drawn the equals sign... but by then, you may have a 2–3 minute head start on your mental calculation.

What I find more interesting are the tricks that don't involve any cheating, but truly allow the performer to do the complex calculations requested. Most of these are things that could be mastered by any reasonably intelligent person with a good memory. One way to double your speed of multiplication, for example, is to memorize your times tables for all two-digit number combinations from 0 to 99. As you may recall, when you do long multiplication, you look at each column of digits and multiply them, using the simple 9×9 table you memorized as a child. Thus, if you memorized the pairs up to 99×99, you could multiply with two columns at once.

Another valid trick is to think ahead about some subclass of problems, and memorize the supporting data you need to come up with a quick answer. An example of this: suppose you tell people you can, in under a minute, instantly estimate the 23rd root of any number up to 44 digits to the nearest integer. Sounds pretty impressive, doesn't it? What your audience hasn't considered is that 99 to the 23rd power has more digits than that – so if you have memorized the 23rd powers of all the integers from 1 to 99, not an unreasonable amount of memorization, you can pretty quickly figure out the proper root by simply looking up the closest numbers in your mental table.

Such quick calculating acts may still be a conversation-starter at parties, but these days are probably not too impressive as performances, given the spread of miniaturized computing devices and wireless communications that make true cheating ridiculously easy. But still, it's fun to think about the various tricks that can enable non-geniuses to carry out seemingly impressive feats of calculation.

Deceptive Digits
From Math Mutation podcast 206

Imagine that you are a crooked corporate manager, and are trying to convince your large financial firm's customers that they own a set of continually growing stocks, when in fact you blew the whole thing investing in math podcasts over a decade ago. You carefully create artificial monthly statements indicating made-up balances and profits, choosing numbers where each digit 1–9 appears as the leading digit about 1/9th of the time, so everything looks random, just like real balances would. You are then shocked when the cops come and arrest you, telling you that the distribution of these leading digits is a key piece of evidence. In fact, due to a bizarre but accurate mathematical rule known as Benford's Law, the first digit should have been 1 about 30 % of the time, with probabilities trailing off until 9 s only appear about 5 % of the time. How could this be? Could the random processes of reality actually favor some digits over others?

This surprising mathematical law was first discovered by American astronomer Simon Newcomb back in 1881, in a pre-automation era when performing advanced computations efficiently required a small book listing tables of logarithms. Newcomb noticed that in his logarithm book, the earlier pages, which covered numbers starting with 1, were much more worn than later ones. In 1938, physicist Frank Benford investigated this in more detail, which is why he got to put his name on the law. He looked at thousands of data sets as diverse as the surface areas of rivers, a large set of molecular weights, 104 physical constants, and all the numbers he could gather from an issue of *Reader's Digest*. He found the results remarkably consistent: a 1 would be the leading digit about 30 % of the time, followed by 2 at about 18 %, and gradually trailing down to about 5 % each for 8 and 9.

While counterintuitive at first, Benford's Law actually makes a lot of sense if you look at a piece of logarithmic graph paper. You probably saw this kind of paper in high school physics class: it has a large interval between 1 and 2, with shrinking intervals as you get up to 9, and then the interval grows again to represent the beginning of the next order of magnitude. If you randomly throw a dart at it, you are much more likely to hit one of the 1–2 intervals than any of the others, with the probabilities diminishing at higher digits in each order of magnitude.

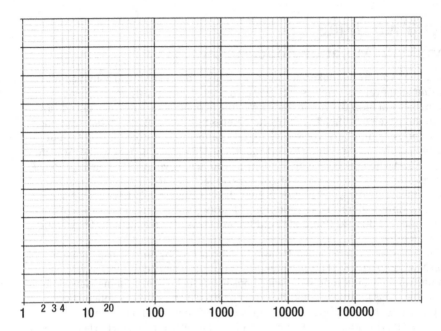

1   2 3 4   10   20   100   1000   10000   100000

Figure 1-1. *Semi-log graph paper*

The idea is that this scale can represent values that may be very small and very large on the same graph, by having the same amount of space on a graph represent much larger intervals as the order of magnitude grows. It effectively transforms exponential intervals to linear ones. If you can generate a data set that tends to vary evenly across orders of magnitude, it is likely to generate numbers which appear at random locations on this log scale – which means that the probabilities of it being in a 1–2 interval are much larger than a 2–3, 3–4, and so on.

Now, you are probably thinking of the next logical question, why would a data set vary smoothly across several orders of magnitude? Actually, there are some very natural ways this could happen. One way is if you are choosing a bunch of totally arbitrary numbers generated from diverse sources, as in the *Reader's Digest* example, or the set of assorted physical constants. Another simple explanation is exponential growth. Take a look, for example, at the powers of 2: 2, 4, 8, 16, 32, 64, 128, etc. You can see that for each count of digits in the number, you only go through a few values before jumping to having more digits, or the next order of magnitude. When you add new digits by doubling values, you will jump up to a larger number that begins with a 1. If you try writing out the first 20 or so powers of 2 and look at the first digits, you will see that we are already not too far off from Benford's Law, with 1 s appearing most commonly in the lead.

Sets of arbitrarily occurring human or natural data that can span multiple orders of magnitude also tend to share this Benford distribution. The key is that you need to choose a data set that does have this kind of span, due to encompassing both very small and very large examples. If you look at a list of populations of towns in England, ranging from the tiniest hovel to London, you will see that it obeys Benford's law. However, if you define "small town" as a town with 1000–9999 residents, creating a category that is restricted to four-digit numbers only, this phenomenon will go away, and the leading digits will likely show a roughly equal distribution.

The most intriguing part of Benford's law is the fact that it leads to several powerful real-life applications. As we alluded to earlier, Benford's Law is legally admissible in cases of accounting fraud, and can often be used to ensnare foolish fraudsters who haven't had the foresight to listen to *Math Mutation*. (Or who are listening too slowly and haven't reached this episode yet.) A link in the show notes goes to an article that demonstrates fraud in several bankrupt U.S. municipalities based on their reported data not conforming to Benford's law. It was claimed that this law proves fraud in Iran's 2009 election data as well, and in the economic data Greece used to enter the Eurozone. It has also been proposed that this could be a good test for detecting scientific fraud in published papers. Naturally, however, once someone knows about Benford's law they can use it to generate their fake data, so compliance with this law doesn't prove the absence of fraud.

So, next time you are looking at a large data set in an accounting table, scientific article, or newspaper story, take a close look at the first digits of all the numbers. If you don't see the digits appearing in the proportions identified by Benford, you may very well be seeing a set of made-up numbers.

Nonrandom Randomness
From Math Mutation podcast 193

Recently my wife got a bit annoyed with me as we drove to a restaurant on our Saturday date night. The problem was that, as usual, I plugged my iPhone into my car's radio to play music for our drive, telling it to shuffle the playlist and select random songs. On this particular night, my phone decided to play four David Bowie songs in a row. Now I should admit that Bowie does take up a nontrivial proportion of my usual playlist, about 160 or so out of the 1000 songs on the list. But it was still pretty surprising that we heard nothing but Bowie on the drive; my wife thought I had set the iPhone on Bowie-only just to drive her nuts. Is such a streak a reasonable result for a truly random song shuffle?

Well, to answer this, we should think about the probability – does it make sense that every once in a while, we would experience a streak like this? It's kind of similar to the well-known "Birthday Paradox". Suppose you are at a party with a bunch of friends, and ask them all their birthdays, looking to see if any two share the same one. You would think that the probability of two people with the same birthday would be pretty low, since if you ask a random person their birthday, the chance of them sharing your birthday is only 1 in 365. But actually, the probability reaches 50 percent as soon as you have 23 people at the party. This seems pretty counterintuitive at first. But think about the number of pairs you have with 23 people: the total number of possible pairs of people is 23 times 22 over 2, or 253. When you look at it this way, the number of pairs seems in the right ballpark to have a decent chance of a shared birthday.

The actual calculation is a bit more complex. An easy way to look at it is by analyzing the party attendees one at a time, and calculating the chances that we do NOT have two people sharing the same birthday. We want to calculate, for each person, the chance that they do not share a birthday with any of the previously analyzed visitors. $P1$, the probability that there are no shared dates yet after looking at the first person, is 1, since there are no people before him. $P2$ is 364/365, the chance that visitor #2 does not have the same birthday as visitor #1. $P3$ is 363/365, the chance that visitor #3 doesn't have his birthday on any of the two days seen so far. And so on. The final probability that nobody has shared birthdays is $P1 \times P2 \times P3 \times \ldots$, up to the number of party attendees. You can see the full probability calculation for this situation at the *Wikipedia* page on the Birthday Paradox. The ultimate result, as I mentioned before, is that if you have 23 attendees, the probability is only about 49 % that there are no shared birthdays.

So, now that we understand the birthday paradox, or at least you're willing to entertain the notion if it hasn't fully sunk in, what does that have to do with shuffled songs? Well, as one author points out at HowStuffWorks.com, you can think about shared artists for songs as something like shared birthdays. My playlist has way fewer artists represented than the number of days in a year, and I have been playing them over and over on many car trips. In the particular case of Bowie, we can see that the odds are better than average, as he represents about 1/6th of my typical playlist. Thus any time a Bowie song plays, there is roughly a 1 in 6 cubed, or 1 in 216, chance it starts a streak of four. And I've gone on a lot more than 216 car rides in my life. So it's not only not unusual, but expected, for me to see regular Bowie streaks. And that doesn't count streaks by other musicians as well, who have slightly lower odds but also are expected to occasionally appear several times in a row.

We also need to keep in mind the human predisposition to look for patterns in randomness. Back when song shuffling first became available on music players, it was a known problem that people would often randomly get the same song twice in a row or only a few songs apart, and then assume something must be wrong with their device. And of course, once people experience this once, they will suffer from a confirmation bias, looking for instances where the same song is repeated and concluding that these verify the supposed technical glitch. Something about our brains just isn't hard-wired to understand or accept the coincidences inherent in randomness. One simple solution was biased random selection, where the device purposely can avoid playing the same artist or song twice in a row based on user settings. Another change that helps is that most current music-playing devices shuffle the music like a deck of cards, creating a full random ordering of all the songs in the playlist, rather than randomizing after each song. This inherently prevents repeats until the user chooses to re-shuffle their list.

To see an extreme case of our human predisposition towards finding patterns, try flipping four coins, writing down the results, and asking a friend if they appear random. No matter what combination you get, it will probably look nonrandom to your friend! If you get 3 or 4 of the same result, such as Heads Tails Heads Heads, it will certainly seem like the coin was biased. If you have 2 of each result, there is no way to avoid having it look like a pattern: either a repeated pair like HTHT, or a symmetric pair like HTTH. You have to really think about it to convince your brain of the randomness of such a set of coin flips.

So, when dealing with birthday sharing at parties, coin flips, music shuffling, or annoyed spouses, remember that sometimes truly random results can seem nonrandom, and try to take a step back and really think about the processes and probabilities involved.

A True Holiday Celebration
From Math Mutation podcast 50

Today I'm going to tell you how to truly celebrate Christmas: you can breathe in one of the actual molecules of air that was in Baby Jesus's first breath! Actually, this form of celebration doesn't take much effort, since you're inherently doing it, whether you planned to or not. But perhaps if you recognize the holy nature of some of your molecules, it will enhance the flavor of the season.

Why do I say you're doing it anyway? Well, think about this. The average human breath contains a number of air molecules roughly equal to 10 to the 22nd power. Now, a baby is pretty small, so for the sake of argument let's cut that down to 10 to the 21st power for Baby Jesus. After two-thousand-plus years, I think it's fair to estimate that those molecules are now pretty well randomly dispersed throughout the atmosphere.

The atmosphere overall contains a number of air molecules approximately equal to 10 to the 44th power. So the proportion of molecules that were actually present in that breath is 10 to the 21st over 10 to the 44th, or about 1 in 10 to the 23rd. But remember that each of your breaths contains an average of 10 to the 22nd air molecules, so you breathe in 10 times 10 to the 22nd, or 10 to the 23rd, air molecules every ten breaths. Thus, in the long term, you expect that on average, you will inhale a Baby Jesus Breath molecule every 10 breaths. Don't you feel closer to Him already?

Incidentally, if you're a non-Christian like me, you don't need to feel left out of the holiday season. This calculation works equally well for Buddha, Moses, Mohammed, Zoroaster, Quetzalcoatl, Akhnaten, or Mitochondrial Eve, assuming each of them actually existed and breathed our atmosphere at some point. I first heard this argument based on Julius Caesar's last breath, in John Allen Paulos' classic book *Innumeracy*, though of course just about anyone who lived that long ago (or longer) is just as good a candidate for this analysis.

Thus, whatever your religious or non-religious tradition, celebrate your holiday season by breathing in deep and sharing some air molecules with your spiritual, biological or philosophical ancestors.

Forgotten Knowledge
From Math Mutation podcast 151

Recently I was reminded of an intriguing science fiction story I read long ago, about a time when pen-and-paper computation had been forgotten. I only remembered it vaguely, but a few minutes of web searching revealed that the story I was thinking of was a 1958 story by Isaac Asimov called "The Feeling of Power".

"The Feeling of Power" is set in a far future, when due to dependence on computers, nobody recalls the basic principles of mathematics, even simple things like pen-and-paper methods of addition and subtraction. Even the basics of how to design computers have been forgotten, since for many generations computers have been designed only by automated programs built into other computers. In this strange world, a gifted technician named Aub discovers that he can perform basic numerical calculations using pencils and paper, using a new technique he calls "graphitics". His discovery is just the common arithmetic we teach today in elementary school. Aub presents his graphitics to a group of government officials, who are amazed at his ability to multiply decimal numbers without having any electronic device available. They are immediately enthralled by the possibilities of replacing expensive computers with cheap human labor throughout the government. In particular, they salivate at the vision of a manned missile, a missile controlled by a human suicide bomber inside, who could make course corrections and do all the needed calculations at a fraction of the cost of a military computer. In the end, Aub commits suicide out of guilt at the death and destruction his discovery will cause.

There are a number of absurdities that stand out in this story. Most glaringly, with the abundant cheap storage and vast supply of information available on the Internet, could basic elements of knowledge like arithmetic ever be truly forgotten? I'm pretty sure that 1000 years from now, even such obscure and useless pieces of data as *Math Mutation* podcast episodes will be easily accessible to anyone who wants them. Asimov's error here is just one small example of the general embarrassment of 20th century science fiction: its overall failure, except in a few isolated cases, to anticipate the Internet and other modern computing technologies.

But there's a more fundamental logical error in Asimov's story that's harder to forgive. If humans could perform pen-and-paper calculations more cheaply and efficiently than computers, why would the human method ever have been forgotten? People replace old tools with cheaper and more efficient ones, not with slower and more expensive ones. Asimov wrote at a time when computers were expensive and gigantic – but even then, in the applications where they were being used, no human armed with a pen & paper could ever compete. ENIAC, the first general-purpose computer, had existed for over a decade before this story, and could perform 5000 operations per second, already far beyond human capacity. And any science fiction writer worth his quatloos should have predicted that computers would only get faster and cheaper from there, until we reached the gigaflops and teraflops, or billions and trillions of operations per second, that we regularly see today.

On the other hand, the story does ring disturbingly true in certain ways. There really has been a trend in recent years to denigrate basic arithmetic skills in our schools. Recently I was frustrated when a fast-food cashier could not properly estimate that two $1.98 hamburgers would cost about 4 dollars. Today the average person knows that in everyday situations like shopping or bill-paying, they can always get out a calculator or run the calculator app on their favorite nearby electronic device. Will the typical person's basic numerical intuition continue to slowly wither away over the generations? We also see the disturbing situation where young Americans today see much more financial reward by going into law or business than science, math, or engineering. It may not be that farfetched that a few generations from now, nobody is alive who remembers how to design a computer chip, as everyone just relies on commoditized manufacturing of existing designs. Even if all the information is documented in theory in Internet archives, how easy will it be to understand for a far-off future generation that has lost their basic experience in engineering?

So, despite its basic flaws, Asimov's story does give us a few things to think about.

Exponents Squared
From Math Mutation podcast 64

Back in school, you probably recall that one of the earliest arithmetic operations you learned was addition. Then the concept was extended to multiplication: take a number x, and add it to itself a bunch of times. A logical further step was exponents: take a number x, and then multiply it by itself a bunch of times. And you probably stopped there – while your math classes went off in various other directions, you didn't really learn any additional operator on this 'ladder'. But why should we stop there? If multiple repeated additions become multiplication, and multiple repeated multiplications become exponents, then what do multiple repeated exponents become?

It might surprise you to learn that there are further operations in this series, though they usually don't bother to teach them in school. After exponentiation comes what is known as "tetration". Tetration is usually symbolized with a number drawn to the upper left of another number, as opposed to the upper right used for exponents. Let's look at an example: what is 2 tetrated to the 4th? We would draw this as a 2 with a little 4 to the upper left side: $^{4}2$. This is equivalent to $2^{2^{2^{2}}}$. Be careful here – we need to start expanding at the uppermost exponent, otherwise we are effectively just multiplying the exponents together. So $^{4}2$ becomes $2^{2^{4}}$, or 2^{16}, which is 65536.

Needless to say, on positive whole numbers, tetration causes values to grow incredibly fast. This probably helps explain why it's not too practical in real life: after all, the simple operation of exponentiation allows us to concisely express the number of atoms in the universe, approximately 10^{80}, or a 1 with 80 zeros. In contrast, if we look at the relatively simple tetration value of $^{3}10$, this is $10^{10^{10}}$, which is 10 to the 10 billionth power, or a 1 with ten billion zeros after it. I couldn't even begin to describe $^{80}10$ in any comprehensible form, other than just saying "10 tetrated to the 80th". It's hard to imagine a real-life application that needs such immense numbers.

But even without real-life applications, there are some mathematicians that find tetration a very interesting topic to study. On small numbers, it can have some bizarre properties. For example, get out a calculator and try calculating some tetrations of the square root of 2. $^{2}\sqrt{2}$, or $\sqrt{2}^{\sqrt{2}}$, is approximately 1.63. $^{3}\sqrt{2}$ is approximately 1.76. And as you increase the tetration, you will find that bizarrely, $\sqrt{2}$ tetrated to higher and higher values always gets closer and closer to 2. So in effect, $\sqrt{2}$ tetrated to infinity is just 2, instead of growing huge like you would expect.

In addition to converging on small numbers, there has also been extensive study into the properties of tetration on complex numbers, which leads to lots of pretty multicolored pictures when graphically plotted on the complex plane. There is actually a web domain, "tetration.org", devoted entirely to the study of this operation, as well as some other sites centered on this topic.

And, as you probably suspected already, tetration is just the beginning. Mathematicians with lots of time on their hands have defined further operations in the series: just as tetration is a chain of exponents, an operation called "pentation" is a chain of tetrations, "hexation" is a chain of pentations, and so on. This general series of definitions is known as the "hyper-operation sequence". As with tetration, none of these have any real-life applications, as far as I can tell with a little web searching. But has that ever stopped mathematicians before?

This whole class of operations, however, is actually a subset of a general operation that is known as the "Ackermann Function", defined by German mathematician Wilhelm Ackermann in the 1920s, which does have some significance in theoretical computer science. So, at some level, these hyper-operations do have a tenuous connection to reality.

Giving You the Fingers
From Math Mutation podcast 75

I remember one day, a number of years ago, when my then-21-month-old daughter came home from preschool, held up her hand, and proudly counted out loud "1-2-3-4-5". My excitement died down a bit when I asked her to count our cat's ears, and she recited the same "1-2-3-4-5". In case you're wondering, our cat is a bit of a mutant, but not *that* much of one, and only has the standard two ears. He does have 12 nipples, but that's a subject for another podcast. Apparently to my daughter, "1-2-3-4-5" was a new word, and is the answer to any question that involves the words "count" or "how many".

But this got me thinking about the process of finger counting. Many of us remember that when we were small, we first learned to count to 10 on our fingers. Is that really an efficient use of finger real estate though? If you think about it, using your fingers to count to 10 is really laboring under a very artificial restriction – that at any time, you can only hold up one finger. If you think of each finger as a binary, or base-2, digit, like a bit in computer memory, then your hands are equivalent to a 10-bit memory. Thus, by holding up various combinations of fingers, you can represent numbers up to 2 to the 10th power –1, or 1023.

This leads to another line of thought: in premodern societies, especially where common merchants were unlikely to have access to cash registers, cheap plentiful scrap paper, or the education to have a good written system of mathematical notation, how did people keep track of numbers? I browsed a bit on the net and found articles on a wide variety of finger-counting methods from different cultures.

In China, a method for representing numbers up to 10 on a single hand has been passed down for many generations. This is based on moving the thumb, index finger, and middle finger into various positions, rather than just up or down. The simpler Korean "Chisenbop" method uses the thumb to represent 5, and the other fingers to represent one each; by holding up multiple fingers, you can again represent 0 through 10 on one hand. Either in the Chinese or Korean system, you can then easily count to 100 on two hands, using one to represent the ones and the other for the tens.

Once you start varying finger positions like the Chinese did, however, this opens up a world of possibilities. Another medieval finger counting system used in Europe and the Middle East uses a combination of finger positions to count up to 100 on each hand: you use your thumb and index finger in ten different positions to represent the tens place digit, while the other three fingers in various positions represent the ones place digit. This means that by using both hands, you can represent any four-digit base 10 number – so you can count up to 9999 on your hands! There is a serious drawback though – it requires a major feat of manual dexterity. Can you hold your pinky and middle finger halfway down, your ring finger straight up, and have your index finger touch the middle of your thumb? That's the number 15 in this system. I tried doing a few more numbers on my hands, and was pretty hopeless. Could you imagine doing a series of these on each hand and holding the position long enough to complete a business transaction?

In any case, can we do better than 10000? I also found an intriguing indirect comment on a yoga site, talking about a system of holding up individual joints or tips of each finger. This means that each non-thumb finger can represent four possible values, and the thumb can represent three, for a total of $4 \times 4 \times 4 \times 4 \times 3$, or 768, combinations on each hand. Thus, with two hands, we can represent a total of 768×768, or 589,824 combinations! Something tells me, though, that not even the most advanced yoga master would ever be capable of this theoretically possible feat.

On the other hand, I can't even count to 10000 on my fingers like a medieval merchant, so who am I to judge what's possible? Perhaps some of you champion videogamers out there can perform feats of manual dexterity that I could never even conceive of.

■ ■ ■

CHAPTER 2

Into the Infinite

What does the word "infinity" mean? This is one of the strangest mathematical ideas we can try to twist our mind around. Does it represent everything in the universe? Is it just a number that is bigger than all other numbers? Can something be greater than infinity? When you try to think carefully about the concept of infinity, all your preconceived notions start to fly out the window. There also seems to be some implied tie-in with religion: how many modern faiths refer to god as "The Infinite" in some form? The idea of multiple infinities was the theme of the very first *Math Mutation* podcast, and there have been many subsequent episodes where we have explored different aspects of this concept.

Too Infinite for Me
From Math Mutation podcasts 0 and 5

Today I want to talk about Hilbert's Hotel, a famous paradox introduced by the mathematician David Hilbert, challenging naive ideas about infinity. Imagine a hotel with an infinite number of rooms, numbered by the natural numbers 1, 2, 3,.... This cannot actually exist, of course, though I hear they are trying to build one next to the Mall of America. Anyway, suppose you arrive at the desk and are told that the hotel is full. Can they still find a room for you?

Surprisingly, the answer is yes, if you can just get a little cooperation from your current guests. The guest in room 1 moves to room 2. The guest in room 2 moves to room 3. And in general, the guest in room n moves to room $n+1$. Since all numbers can be increased, everyone still has a room, and you get one too!

Now suppose you happen to get married to a fellow math geek, try out the latest and greatest fertility drugs, and nine months later your wife bears a set of infinituplets, with one child for each natural number. Once they are old enough, you and your infinite family decide to go on vacation to Hilbert's Hotel, and arrive at the front desk demanding enough rooms to accommodate all of you. Can they still manage this? Yes, they can. The guest in room 1 moves to room 2. The guest in room 2 moves to room 4. And the guest in room n moves to room $2n$. Every number can be doubled, so everyone still has a room. But now all the odd-numbered rooms, of which there are infinitely many, are free for your family to take up.

What we are seeing here is that in some sense, all 'countably' infinite sets are equivalent – even when some seem to be subsets of others. When moving guests to different rooms in Hilbert's Hotel, we were effectively setting up mappings between the

© Erik Seligman 2016
E. Seligman, *Math Mutation Classics*, DOI 10.1007/978-1-4842-1892-1_2

set of current rooms, the integers from 1 to *n*, and other sets: the set of integers greater than 1, or the set of even integers. Using these mappings, each of these sets can be fully accommodated in new rooms of Hilbert's Hotel, while leaving some other rooms free.

Now let's ask another follow-up question: is it possible that such a large number of guests could arrive that they could NOT fit in the hotel? This is another case where the surprising answer is yes.

Suppose those fertility drugs were more effective than you expected, and your wife has given birth to a very special set of infinituplets. This is no garden-variety, boring, everyday set of infinituplets – you have so many children that you have one for every non-negative real number strictly less than 1. One child is numbered .11111 (repeating), another is .12345, another is a decimal followed by all the digits of π, etc. To keep track of them all, you give each child a T-shirt with a different real number on it (represented without an infinite tail of 9 s, to avoid technical issues).

Now, arriving at the hotel, you decide you want each child to have a room of their own. You tell your children to enter in an orderly line, and you write some things down as they walk in. Start by writing a decimal point on your paper. When your first child is assigned room 1, you write down the *first* digit from his T-shirt. For the child assigned to room 2, you write down the *second* digit, and so on. Once all the rooms in the Hotel are full, you have written down a real number on your paper, such that its first digit matches the first digit from the child in room 1, the second digit matches the second digit from the child in room 2, and so on.

Next, under the number you wrote, you write another number as follows: if the digit in the original number is 8 or 9 you write 0 underneath it; otherwise you add one to the digit and write that digit underneath the original (so, for example, under the number .2859328... you would write .3060430...). This number cannot match *any* of the numbers of the kids in the rooms - its first digit differs from the first digit in room 1, its second digit differs from the second digit in room 2, and in general its *nth* digit differs from the *nth* digit in room *n*, for *any* value of n.

But we said you had a child for every non-negative real number less than one! So, the child with the number you just wrote down is sitting in the lobby, crying that you ignored him. Maybe you should buy him an ice cream cone to calm him down. I hope you brought a lot of spare change though, since an infinite number of his brothers and sisters, whose numbers can be discovered through variants of the same method, were also left behind in the lobby.

This is essentially the famous 'diagonal' proof discovered by Georg Cantor in 1891, showing that not all infinities are equal. The *uncountable* infinity of real numbers is strictly larger than the *countable* infinity of positive integers: no matter how you assign corresponding integers to each real number, after you assign all the integers, some real numbers will be left over.

Infinitely Ahead of His Time
From Math Mutation podcast 116

As mentioned above, 19th century mathematician Georg Cantor was the founder of modern set theory, and proved many surprising results dealing with infinity, such as the proof that the infinity of real numbers is greater than the infinity of integers. These results are now well-accepted and seen as part of the foundations of modern mathematics; if you attended a U.S. elementary school in the 1960s, you may even recall the ill-fated "New Math" fad that tried to teach set theory to children. But it might surprise you to hear that in his own time, Cantor's theories were seen by many as incorrect, dangerous, or blasphemous. And oddly enough, he found support from some religious Catholics while his ideas were being rejected by prominent figures in the math world.

To start with, let's review Cantor's most controversial theories. Before his time, infinity was seen as a kind of vague abstraction, not something vulnerable to formal types of analysis. When Cantor developed set theory, he defined two sets to have the same 'cardinality', or number of elements, when you can set their elements in 1-1 correspondence. Once he had defined this, it seemed reasonable to apply the definition to infinite sets as well. Looking at the infinite set of natural numbers, 1, 2, 3, etc, and the infinite set of real numbers, i.e. all decimals, he asked if these two infinite sets had the same cardinality. With his famous diagonal proof, he showed that no matter how you try to set up a correspondence between natural numbers and real numbers, some real numbers would be left out. Thus, even though both sets are infinite, the infinity of natural numbers is somehow smaller than the infinity of real numbers. And extending this idea, for every infinite set he could define a larger infinite 'power set', determining an infinite hierarchy of infinities.

In the mathematical world, there were three major groups opposed to Cantor's ideas. The *constructivists*, including Cantor's former mentor Leopold Kronecker, believed that a mathematical object could not exist without an explicit method to construct it. Since any attempt to construct an infinite set is non-terminating, this generally invalidated results dealing with infinity. Kronecker even labelled Cantor a "corrupter of youth" for teaching his odd theories to his students. The *intuitionists*, such as Henri Poincaré, had a related but slightly different objection: they considered mathematics a tool for more deeply analyzing intuitive notions of the human mind. An uncountable infinite set, such as the real numbers, is seen as beyond human intuition and thus not mathematically analyzable. The *finitists* were a stricter offshoot of these two schools, believing that no mathematical object could exist unless constructible from natural numbers in a finite number of steps.

On the religious side, many Christian theologians were horrified at the concept of multiple types of infinity, since they only recognized one true infinity, God. Cantor made a concerted effort to bring the religious community to his side, however: he believed that his multiple infinities increased God's glory, considering God to be an absolute infinity above the hierarchy of mathematical infinities he had discovered. Indeed, didn't it make God even more powerful, if there were below him not only finite concepts, but an infinite set of lesser infinities? He wrote to numerous prominent figures, including the Pope. While his efforts had mixed results, he did win over a few well-known religious philosophers, such as Catholic priest Constantin Gutberlet, who believed Cantor's theories of infinities provided a tool for man to probe more deeply into the nature of the Divine. In an odd twist of historical fate, he seemed to have more support at times from the Catholic Church than from the scientific community.

All the controversy over his ideas took quite a toll on Cantor, causing lifelong issues with depression, though some modern scholarship suggests that his problems may also have been a result of undiagnosed bipolar disorder. After being promoted to a full professor at age 34 and publishing a series of seminal papers that formed the foundations of set theory, Cantor was requested to withdraw a paper from the prestigious journal *Acta Mathematica* in 1884. He was soon hospitalized for depression, and dropped out of mathematics completely for a while. Oddly, he decided to spend some time researching the works of Shakespeare, and published several pamphlets suggesting that Francis Bacon really wrote the plays. A few years later he seemingly recovered and returned to his profession, but suffered a similar bout of depression in 1899, and afterwards these continued every few years. He remained active in the mathematical world, but his work continued to be controversial. He eventually died in poverty in 1918.

However, towards the end of his life Cantor's mathematics was increasingly accepted, and he received numerous prestigious honors in his lifetime. In 1904, the Royal Society in London granted him the Sylvester Medal, its highest award. And in 1911, he was one of the distinguished foreign scholars invited to the 500th anniversary of the founding of St Andrews University in Scotland. David Hilbert, a leader of 20th century mathematics, summed up Cantor's work with the statement: "No one shall expel us from the Paradise that Cantor has created."

Infinite Infinities
From Math Mutation podcast 157

During a recent evening, I told my daughter that it was 10 minutes until bedtime. She replied, "No, infinity minutes!" Naturally, I responded, "Are you talking about a countable infinity, analogous to the integers, or the uncountable infinity of the real numbers?" Since she stared at me blankly without a response, I concluded that we would have to stick with the original plan of 10 minutes.

But afterwards I realized that my question hadn't been quite valid. Rather than just two classes, there are actually an infinite number of ever increasing infinities. It's pretty easy to prove, based on a variant of Cantor's famous diagonal argument, the same one we used to prove there are more reals than integers.

The key is to recognize that for every set there is a "power set", the set of all subsets of its elements, that is strictly larger. For example, look at the set consisting of elements a, b, and c. The power set contains the subsets {a,b}, {b,c}, and {a,c}, plus the trivial singleton subsets {a}, {b}, {c}, the null set, and the full set {a,b,c}. For a finite set, we can see that the power set is obviously larger than the set we started with.

But is it the case that the power set of an infinite set is larger than the original set? At first this might seem obvious as well: since all the singleton sets are trivially in the power set, it means that the power set has all the elements of the original set plus a bunch of other stuff. However, this kind of reasoning doesn't work with infinities. A simple counterexample is given by comparing the set of whole numbers to the set of even numbers, as we discussed earlier. At first you would think there are twice as many whole numbers, since it contains all the even numbers plus the odd numbers. But by assigning every even number 2 N to room N in Hilbert's Hotel, we can see that the full set of even numbers precisely fits. Thus there is a 1-1 correspondence between even and whole numbers, and their infinities are actually equivalent.

So, to show that for any set S, its power set P is truly larger, we need to show that in any attempt to set up a 1-1 correspondence, we will have some excess element of P that doesn't fit. Let's look again at a Hilbert Hotel, and assume it has a room numbered for every element of P. Can we show that after guests arrive with every element of S on their shirts, and each is assigned a room, there will always be a room left over?

Let's assume on the contrary that we can fill the hotel, so that every room is occupied by a guest. So for every guest labeled with some element of S, he is assigned a room numbered with an element of the power set P. Remember that the power set is just the set of subsets of S, so each room will have a list of guests on its door, each of whom may or may not be in that room. Let's define another set Q as the set of guests who are in a room that does not include their label on the door. (Q is not empty since there is a guest in the room labeled by the null set.) Since Q is a subset of S, there must be a room labeled with the set Q on its door.

Now, by assumption all the rooms are occupied, so let's look at the guest assigned to room Q. If his label is on the door, we have a contradiction, since we defined Q as the set of labels of guests who are NOT in rooms with their label on the door. If his label is not in Q, then this guest is in a room without his label on the door – which means he must be in Q, due to the way we defined it. Thus we have another contradiction. So, no guest can be in room Q, and room Q must be empty, contradicting our assumption, and thus, the power set is strictly larger than the original set. And since every set, including a power set itself, has a power set that is strictly larger, this proof shows that there is an infinite class of infinities.

So, next time your child wants to delay their bedtime, you'll be prepared to discuss in detail which class of infinite extensions they are requesting.

Infinity Times Infinity
From Math Mutation podcast 184

If you were watching American TV around 2012–2013, you probably saw the silly commercial where a TV announcer is interviewing a bunch of kids, and asks them to name the largest number they can think of. One says "infinity", and the next one tops him by yelling out "infinity + 1". The announcer then says, "Sorry, I was looking for infinity + infinity." But then a little girl tops him, by blurting out, "What about infinity *times* infinity?" I'm inferring that the viewer is supposed to be amused by all these meaningless manipulations of infinity that don't really make much sense. But did you know that there is a well-defined extension of the real numbers that includes clear definitions of operations on infinity, and that it is actually possible to intelligently compare infinity + 1, infinity + infinity, and infinity times infinity?

This set of numbers is known as the "surreal numbers", and was first defined (or discovered, depending on how you look at it) by John Conway, and then popularized in a book by Donald Knuth. By the way, I should take a moment to congratulate Professor Conway on his recent retirement from the Princeton University Mathematics Department – somehow his name always seems to come up in conjunction with the cool math stuff we talk about here, and I have the feeling we haven't heard the last from him. The definition of surreal numbers is kind of unusual, but there are good reasons for this. Of course it's pretty easy to define any arbitrary stuff and call it infinity, without really understanding what you're talking about. But it's challenging to define a system of numbers that is self-consistent, encompasses the reals we already know, adds infinities, and behaves as expected under addition, multiplication, and similar operations.

The definition of surreal numbers works like this: a surreal number is a pair of sets, a left set and a right set, such that every element of the left set is less than every element of the right set. In the simple cases, such a pair can be thought to represent a value intermediate between the two sets. So, for example, the surreal number equivalent to 1/2 can be represented by a left set of 0 and a right set of 1. Any real number can be represented by an infinite left set consisting of all rationals less than it, and an infinite right set consisting of all rational values greater than it. The advantage of this set-based definition is that it's then possible to clearly define all the basic arithmetic operations in terms of these left and right sets, and show that such operations are closed: if you start with surreal numbers, the result will be a valid surreal number as well. The details are a little messy, but you can find them in the more detailed sources in the References chapter.

At this point you might ask, what does all this have to do with infinity? Well, remember that the left and right sets can be infinite sets – so there is a surreal number defined by the left set of all integers, and an empty right set. This is an infinite number, larger than any integer, traditionally represented by the Greek letter ω (omega). Then, after that, we can define $ω + 1$ as having a left set of ω and an empty right set: this is a number greater than ω. Similarly, $ω + 2$ is greater than $ω + 1$, and so on... until eventually we can define $ω + ω$. Going on to multiplication, we can define a number ω times 2, whose left set contains $(ω + n)$ for all n. We can continue to define multiples of ω, until we reach ω times ω, just like in the commercial. Clearly that kindergartner had an advanced knowledge of set theory. But there is no reason she had to stop at ω times ω. ω can be raised to arbitrary exponential powers as well, to get even larger numbers, such as $ω^ω$.

There are other manipulations of infinity, represented as ω, that are also pretty cool in this system. Integers can be subtracted from ω, so that $ω - n$, for any natural n, is less than ω, but still larger than all integers. All such numbers are greater than the smaller but still-infinite value $ω / 2$. ω also has an inverse, ε (epsilon), the infinitesimal number: ε is greater than 0, but less than any other real number. ω is equal to $1/ε$, so as you would expect, ε times ω equals one. You can also perform various operations on ε, similar to what can be done with ω, to find other infinitesimal numbers that might be greater or smaller, while still lying between 0 and all the familiar reals on the surreal number line.

We should probably also take a minute to clarify that there was an earlier definition of ω before the surreal numbers came about, by Georg Cantor in the 19th century, as part of his system to extend the system of ordinal numbers to accommodate infinities. In Cantor's system, ω is defined as the smallest infinite number denoting the size of a well-ordered set. His definition of the arithmetic operations on ω had a few peculiarities, such as addition and multiplication not being commutative: $1 + ω$ is just ω, but $ω + 1$ is a larger number. This kind of makes sense in some ways: if you are performing the operation "go all the way to the end of the infinite number line," which is what adding ω is like, it doesn't matter that you started a few steps ahead, you're going to the same place. But if you're already at the end of the infinite line, then you can go farther, if that doesn't make your brain hurt. Anyway, the surreal numbers are said by some to be more satisfying due to their more familiar arithmetical properties, which do match our standard expectations: in surreal numbers, $1 + ω$ equals $ω + 1$.

So, is there any application of the surreal number system, besides describing the bizarre variety and properties of these infinite numbers, and inspiring quirky television commercials? Strangely enough, Conway first came up with surreal numbers when trying to come up with a mathematical analysis of the game of Go. The definitions of infinite constants were not really his intention, but were side effects of his investigations into game theory. I think the big lesson here is that, if you think hard enough and are careful about your definitions, concepts like arithmetic operations on infinities, which may seem too bizarre or nebulous to define, might turn out to make sense after all. Even to a group of 5-year-olds.

Infinite Perimeter, Finite Area
From Math Mutation Podcast 21

Suppose I asked you to draw me a figure that has an infinite perimeter, but a finite area. Could you do it? Well, not literally, since your wrist would get kind of tired trying to draw an infinite perimeter. But it might surprise you to know that we actually can describe such figures. One famous example is known as the Koch Snowflake or Koch Star. It was first described by mathematician Helge von Koch in 1904.

To define this figure, start with an ordinary equilateral triangle. Then you want to draw a spike jutting out from each side. More precisely, imagine the center third of each side as the base of a smaller equilateral triangle, and draw your spike by drawing the other two sides of the triangle, erasing the original segment. After you have done this for all three sides, the figure will look like a six-pointed star. Now repeat the process, drawing similar spikes in the middle third of each of the twelve segments making up the star.

Continue for an infinite number of steps, each time drawing spikes out of the center third of all line segments in the figure. The result, looking somewhat like an infinitely complex snowflake, is the Koch Star:

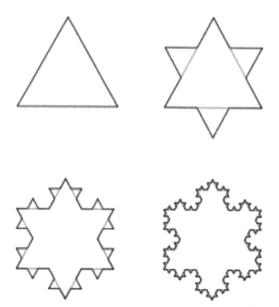

Figure 2-1. *First steps in building the Koch snowflake[1]*

---

[1]Image used with permission under the Creative Commons Attribution-Share Alike 3.0 Unported license; original found here: https://commons.wikimedia.org/wiki/File:KochFlake.svg. Copyright user Wxs.

Does this really have an infinite perimeter? Well, the easiest way to see this is that each step takes every individual line segment, and multiplies its total length by 4/3, since we are replacing the middle third with two segments of equal length. Multiplying the original perimeter by something greater than 1 an infinite number of times will, naturally, get us an infinite result.

Does it really have a finite area? Despite its infinite perimeter, the answer is yes. The simplest way to see this is probably to draw a circle that just barely encloses the six-pointed star you generate after the first step, touching it at each of the six points. If you sketch out the next few steps of drawing spikes for the Koch Star, you will see that none of the spikes can ever jut outside that original circle. Thus, however long you continue drawing spikes, the total area must be less than or equal to the area of that circle.

So, what is the significance of this figure? Aside from the fact that it looks cool, it is one of the simplest examples of the class of figures known as fractals. Fractals are interesting because they exhibit greater detail the closer you examine them. For example, suppose I show you a picture of a Koch star. You might spot one small portion that looks smooth. But if you examine it more closely with a magnifying glass, you will see more and more complexity, inherent because any line segment in the figure must, by definition, have a spike sticking out of its center.

These kinds of figures can be used as models for real-life phenomena like mountains and coastlines. For example, a coastline looks smooth when viewed from an airplane overhead. But if you take a walk on the beach, you will see much more jagged detail than from the high-level view. And if you crawl along the beach examining grains of sand with a microscope, not only will you look like a dork, but you will also realize there is an even greater level of detail than you noticed when walking along. Due to their ability to model this type of complexity, fractals have applications in geology, seismology, medicine, art, and various other areas.

A Pretty Big Library
From Math Mutation podcast 15

Wouldn't it be nice if you could go into a library and be guaranteed that any possible book would always be there? I'm not just talking about books that have been written – this library contains all books that could ever be written. If we assume that any books over a fixed length, say 410 pages, are split into multiple volumes, the library wouldn't even have to be infinite. Let's say that each page contains 40 lines, and each line contains 80 characters. That's 3200 characters per page; multiply by the 410 pages, and that's just 1,312,000 characters per book. If we assume that each character can be one of 26 letters, a space, a period, a comma, or a quote, then each character has 30 possibilities. Then the total number of possible books is a mere 30 to the 1,312,000th power. Of course, since modern physics claims that there are only about 10 to the 80th power atoms in the observable universe right now, I'm not quite sure where we would build this library. In fact, with a new storage technology that can represent an entire book on one atom, you couldn't even build a virtual version of this library. But it's still fun to speculate about.

There would be quite a few implications of this very large library. To start with, as I mentioned, it contains every possible book. Are you an aspiring author? Well, you'd better pick another profession, since all the books you will ever write are already on the shelf. You can even find one volume that contains transcripts of all the *Math Mutation* podcasts, including the ones that I haven't recorded yet. But a slight twist to this is that every possible variant of each book is also there – so there is a fake volume that seems to contain all *Math Mutation* podcasts, but where I get the math entirely wrong in every one. (At least, I hope that one's a fake volume.) Finding the volume you want would be kind of hard though.

Another implication is that no matter what your problem is in life, no matter what your goals or dilemmas are, the answer is in one of the books. There is a complete book of prophecies, describing the origins of man and outlining all the major events of history, until our race's downfall. Likewise, there are many books of heresies, describing a false past and steering us in every possible bad direction. There are also plenty of books of bad advice, including a complete collection of Dr. Phil's writings. And again, finding what you want would be a challenge.

But one possible multi-volume set that exists is surely an *index* to the library, essentially a multi-volume card catalog; so maybe you can find the volume you want! Given the large portion of the books that are pure gibberish, finding this would probably be our first priority anyway. Someone just has to find the index, and then you can get to whatever volume you need. On the other hand, there are also plenty of fake indices that are right the first few times you try them, but after that send you to a wrong shelf light-years out of your way. There's even an index that is guaranteed to send you to a slightly wrong version of any book, where the last chapter is replaced with a series of Yoko Ono lyrics. So I guess we're hosed anyway.

As you have probably guessed, this idea is not original with me – it comes from the classic short story "The Library of Babel", by Jorge Luis Borges. I highly recommend his *Collected Fictions*, which touch on a lot of ideas like this on the border of mathematics and the humanities.

Someone Knocked My 8 Over
From Math Mutation podcast 28

The ∞ symbol for infinity, which looks like the numeral 8 on its side, has always seemed kind of weird to me. Perhaps that is fitting, since infinity is certainly a weird number, as we have discussed. I had to stare at it for a while to figure out what specifically bothered me about the symbol. One thing is that it's kind of short and fat, compared to most other symbols in our math repertoire. Another is that it represents a quantity, sort of, but isn't a letter in some language, like most constants such as π or 'e'. The symbol was first used explicitly to represent infinity in a book by English mathematician John Wallis in 1655, but he did not supply any details on why he used this symbol. So where does the infinity symbol come from?

I like Rudy Rucker's explanation, in his classic book, *Infinity and the Mind*, that it represents a Demolition Derby racetrack, where the cars can zoom around forever. But I suspect Wallis didn't attend too many car races in 1655, so there is probably a better explanation.

One common thought is that maybe the symbol represents a Möbius strip. As you may recall, this is a strip of paper that has its ends taped together after a half-twist: if you do this, any line you start drawing along the paper will eventually appear on both sides without you ever having to lift your pen. This does suggest ideas of infinity. And held at the right angle, the strip does look kind of like an infinity symbol. Unfortunately, the Möbius strip was discovered in 1858, so Wallis would not have known any more about that than about demolition derbies.

Other theories of its origin refer to ancient religious symbols, such as the "Ouroboros", a drawing of a reptile or dragon eating its own tail. Some representations of the Ouroboros are drawn in a way that does look like the infinity symbol. There are Ouroboros-like symbols or references in such diverse areas as Plato's writings, Christianity, Aztec ruins, and African religions.

It is also possible that Wallis based his infinity symbol on slight modifications of Greek or Roman symbols. Perhaps it is simply a perversion of the letter omega, the final letter of the Greek alphabet, which looks like a curvy 'w'. Or maybe it came from an Etruscan and Roman symbol for 1000, which was also sometimes used to indicate large abstract quantities, and looked like a 'C', followed by a vertical line, and then a backwards 'C'. I guess the vertical line acted like a twisty-tie to pull in the edges of the C's, and keep the infinite garbage from spilling out of them.

One more explanation that seems kind of fun is that the symbol represents an hourglass turned on its side. If you think about it, when an hourglass is on its side, no time will seem to pass, so any interval being measured will literally seem to extend to infinity. Or at least until the sun goes nova and melts the hourglass; but that's still a pretty long time, if we can trust our astronomers.

Finally, there is the mundane explanation that in the days of manual typesetting, creating a new symbol was a pain, so Wallis just used an 8 turned on its side to simplify the printing process. I think this is my favorite explanation, just because it makes all the people pondering this question look really silly, and making people look silly is always fun. Of course, after this podcast, I guess I may be one of them. But we will never really know.

Not Quite Infinity
From Math Mutation podcast 186

In one of the previous sections, we talked about a group of kindergarteners challenged to name the largest number, eventually settling on infinity times infinity as the winner. But what if they had made a rule against infinities? As our more astute readers are probably aware, that would be a silly rule, because the whole point of infinity is the fact that numbers go on forever, so regardless of the current answer, you could always name a bigger non-infinite number. But what if the question is about the largest number used somewhere in a serious mathematical proof? Then it becomes a little more interesting. In that case, it would not be surprising if one of the children had named Graham's Number, popularized by one of Martin Gardner's *Scientific American* columns in 1977, which was listed for many years in the *Guinness Book of World Records* as the largest such number.

What is Graham's number? It arises from a discussion by American mathematician Ron Graham about connecting all the vertices of an n-dimensional hypercube and coloring the edges. What is the smallest number n for which coloring the edges of such a hypercube-based graph, with 2 colors, guarantees at least one single-colored complete subgraph of 4 vertices? You don't really need to understand the problem to hear about the number though.

To describe the number, we need to start by describing what's called Knuth's up-↑ notation. You can think of this as a generalization of the concept of exponents. To start with, "a ↑ b" is the same as a^b, the result of repeated multiplication. So 4 ↑ 3 is $4 \times 4 \times 4$, or 64. Then, each time you add an ↑, it's a result of repeated operation of the previous type of ↑. So 4 ↑↑ 3 means 4 tetrated by 3, or 4 to the 4 to the 4th power, or 4^{256}. This is already getting pretty big, over 150 digits when written as a decimal. Similarly 4↑↑↑ 3 means 4 tetrated by 4 tetrated by 4, an even more astronomical value.

Now that we have this arrow notation, we can define Graham's actual number. Define $g1$ as 3↑↑↑↑ 3. Note that this number is already almost inconceivably immense. Then $g2$ is defined as $3\,g1$-↑ 3. In other words, in Knuth's ↑ notation, the number of ↑s used to define $g2$ is equal to the full value of $g1$. Repeat this for 64 steps, with g_n equal to $3\,g_{(n-1)}$-↑ 3. The final number G, Graham's number, is equal to g_{64}. The *Wikipedia* page points out that if you tried to write this final number in ordinary decimal notation, the observable universe would be too small to contain it, even assuming your handwriting is small enough that each digit fits in a Planck volume, around 10 to the minus 105 cubic meters.

After Gardner's publication, and its subsequent appearance in the *Guinness Book of World Records*, this number caught on for a while in pop culture; one post I spotted online talks about a mathematician who was asked about it by a cab driver. Ironically, Graham's Number isn't the real number defined for use in Graham's published proof. Later on, Graham admitted to a reporter that his famous number was actually a weaker bound than the true one in the theorem, slightly larger than necessary, because the real one was too hard to quickly describe to a non-specialist. We also shouldn't treat the *Guinness Book of World Records* as a very good source of mathematical wisdom; while they have removed the Graham's number record from their online edition, they describe the 'smallest infinity' as the 'sum of all the integers'; not a very good description, even if arguably equal in magnitude to Cantor's definition. It's also the case that since then, several larger numbers have been generated by other unrelated proofs, though somehow none seem to have caught on in pop culture.

What amuses me most about Graham's Number is the sense of intimidating magnitude you get when thinking about a number so large that, if written down on paper, the paper would fill the entire known universe. Somehow in podcasts like this we have gotten almost blasé about discussing infinities, even though every infinity is technically larger than Graham's number, and don't get the same sense of awe. Maybe we need to think about large numbers like this every once in a while, to remind us how infinite infinity really is.

Big Numbers Upside Down
From Math Mutation podcast 203

When it comes to understanding big numbers, our universe just isn't very cooperative. Of course, this statement depends a bit on your definition of the word "big". The age of the universe is a barely noticeable 14 billion years, or 1.4×10^{10} years. The radius of the observable universe is estimated as 46 billion light years, around 4.6×10^{25} meters. The observable universe is estimated to contain a number of atoms equal to about 10^{80}, or a 1 followed by 80 zeros. Now you might say that some of these numbers are pretty big, by your judgement. But still, these seem pretty pathetic to me, with none of their exponents even containing exponents. It's fairly easy to write down a number that's larger than any of these without much effort, as we have discussed in the previous section. While it's easy to come up with mathematical definitions of numbers much larger than these, is there some way we can relate even larger numbers to physical realities? Internet author Robert Munafo has a great web page, linked in the show notes, with all kinds of examples of significant large numbers.

There are some borderline examples of large numbers that result from various forms of games and amusements. For example, the number of possible chess games is estimated as $10^{10^{50}}$. Similarly, if playing the "four 4 s" game on a calculator, trying to get the largest number you can with four 4 s, you can reach $4^{4^{4^4}}$, another truly immense value. It can be argued, however, that numbers that result from games, artificial exercises created by humans for their amusement, really should not count as physical numbers. These might more accurately be considered another form of mathematical construct.

At a more physical level, some scientists have come up with some pretty wild sounding numbers based on assumptions about what goes on in the multiverse, beyond what humans could directly observe, even in theory. These are extremely speculative, of course, and largely border on science fiction, though based at some level in modern physics. For example, one estimate is that there are likely $10^{10^{82}}$ universes existing in our multiverse, though this calculation varies widely depending on initial assumptions. In an even stranger calculation, physicist Max Tegmark has estimated that if the universe is infinite and random, then there is likely another identical copy of our observable universe within $10^{10^{115}}$ meters. Munafo's page contains many more examples of such estimates from physics.

My favorite class of these large "physical" numbers is the use of probabilities, as discussed by Richard Crandall in his classic 1997 *Scientific American* article [Cra97]. There are many things that can physically happen whose infinitesimal odds dwarf the numbers involved in any physical measurement we can make of the universe. Naturally, due to their infinitesimal probabilities, these things are almost certain never to actually happen, so some might argue that they are just as theoretical as artificial mathematical constructions. But I still find them a bit more satisfying. For example, a parrot would have odds of about a 1 in $10^{3000000}$ of pecking out a classic Sherlock Holmes novel, if placed in front of a typewriter for a year. Taking on an even more unlikely event, what is the probability that a full beer can on a flat, motionless table will suddenly flip onto its side for no observable reason sometime in the next year? Crandall estimates this as 1 in $10^{10^{33}}$. In the same neighborhood is the chance of a mouse surviving a week on the surface of the sun, due to random fluctuations that locally create a comfortable temperature and atmosphere: 1 in $10^{10^{42}}$. Similarly, your odds of suddenly being randomly and spontaneously teleported to Mars are $10^{10^{51}}$ power to 1. Sorry, Edgar Rice Burroughs.

So, it looks like tiny probabilities might be the best way to envision the vastness of truly large numbers, and escape from the limitations of our universe's puny 10^{80} number of atoms. If you aren't spontaneously teleported to Mars, maybe you can think of even more cool examples of large numbers involved in tiny probabilities that apply to our physical world.

CHAPTER 3

■ ■ ■

Getting Geometric

Ever since Euclid, generations of learners have been astonished by the level of insight we can gain into the basic shapes of the world around us, starting with a few observations and basic principles. Isn't it amazing that you can draw any triangle you want on a flat piece of paper, not telling me anything else about it, and yet I will know that the sum of its angles is precisely 180 degrees? You probably learned that fact in school, but unless you were very lucky or had an especially talented teacher, your textbook sucked all the joy out of it. Here we focus on a small set of geometry-related topics that I still enjoy, decades after my high school class on the topic.

Something Euclid Missed
From Math Mutation podcast 171

Here's an experiment for you. Take a piece of paper, and try using your abundant artistic skills to draw a triangle. Any triangle will do: it can be equilateral, isosceles, right, or none of the above, just the lines connecting three random dots on your paper. Now trisect each angle: draw lines 1/3 and 2/3 of the way across the angle at each corner. There should be three points near the middle of the triangle where pairs of adjacent trisectors intersect each other. Join these to form a little triangle in the center. Assuming you are sufficiently talented to be able to draw straight lines and approximately trisect angles by hand, you will notice something amazing: no matter what triangle you started with, the small one in the middle is equilateral, with all sides the same length and all angles at 60 degrees! How did this happen?

© Erik Seligman 2016
E. Seligman, *Math Mutation Classics*, DOI 10.1007/978-1-4842-1892-1_3

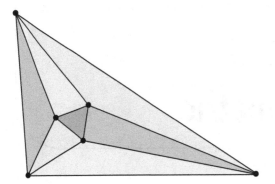

Figure 3-1. *The Morley triangle[1]*

This little equilateral triangle is known as Morley's Triangle. You would think that such a simple trick, drawing lines within a triangle to get cool shape in the middle, would have been part of the classical geometry known since Euclid, but surprisingly, that's not the case. Euclid and his contemporaries may have missed this due to his tendency to concentrate on figures that could be constructed with compass and straightedge, since trisecting angles isn't directly possible with this type of technique. This triangle wasn't discovered until 1899, by Anglo-American mathematician Frank Morley at Haverford College. Morley was actually investigating complex properties of more general algebraic curves, and came across this triangle by accident. He didn't bother publishing it right away, though it spread by word-of-mouth until it eventually appeared in print as a problem in *The Educational Times* in 1908. He also showed it to his young son, who was fascinated by this magic triangle and later reminisced, "Always, to the eye at least, the theorem, if drawn accurately, proved itself. What caused me considerable annoyance was that I could not for a long time comprehend what purblind examiners might accept as a valid proof." These recollections also hint that part of Morley's reluctance to publish may have come from the fact that the theorem seemed so simple and obvious (once drawn) that he was sure somebody must have already discovered it centuries ago. But the first actual publication of a proof was by two other mathematicians named Taylor and Marr in 1913, who acknowledged Morley in their paper.

Since Morley, numerous proofs have been discovered of the theorem. Trying to guess at the intuition behind the Morley triangle, it occurred to me that 180 degrees is a special quantity for triangles, the sum of their angles, so it only makes sense that when messing around with trisected, or 1/3, angles, the quantity 60 degrees, which matches the angles at the corner of an equilateral triangle, would play a special role. Unfortunately, I haven't been able to find a proof that really connects to this intuition as to why this magical equilateral triangle appears. Many proofs are basically solving a set of trigonometric equations to figure out the relations of the lines and angles, fully valid and convincing but not providing much insight. Probably the cleanest proof I've seen online is one discovered in the late 20th century by Conway, where he basically assembles a bunch

[1]Image used with permission under the Creative Commons Attribution-Share Alike 3.0 Unported license; original found here: https://commons.wikimedia.org/wiki/File:Morley_triangle.svg. Copyright user Hagman.

of small triangles with the right sides and angles, and shows how they fit together to form any larger triangle with an equilateral Morley triangle in the center.

But an even more surprising aspect of Morley's theorem is that it can be generalized to find other implied equilateral triangles lurking around. We've been talking about trisecting the interior angles of a triangle – but what about the *exterior* ones? Actually, if you draw the exterior trisectors of each angle of the triangle, you can come up with yet more equilateral triangles, both by intersecting the exterior trisectors with each other, and intersecting interior and exterior trisectors. You can also come up with slightly different trisectors by adding 360 or 720 degrees to the size of an angle and then dividing by three, yielding yet more implied triangles. There are a total of 18 Morley triangles that can be constructed. One amusing article I spotted on the web was from a math enthusiast who wrote a computer program trying to illustrate the central Morley triangle we started with, but due to a bug actually trisected the exterior angles in some cases... and was surprised to produce equilateral Morley triangles anyway!

I think the coolest aspect of this whole Morley triangle concept is that we had a supposedly well-explored, solidly understood area of mathematics, the Euclidean geometry of planar triangles, and thousands of years later a new and unknown property was discovered. Just draw the trisectors of each corner of any triangle, and their points of intersection determine an equilateral triangle. The ancient Greeks could have discovered the Morley triangle and come up with a proof like Conway's, but somehow they didn't, despite some of them having a literally religious devotion to geometry. How many more surprises are lurking in what we today consider well-understood areas of math? Maybe someday a *Math Mutation* listener will be the one who discovers something new. Maybe even you.

How Not to Decorate Your Bathroom
From Math Mutation podcast 82

There's one men's room at the building where I used to work that drove me nuts. Like most corporate bathrooms, the floor consisted of tiles of several different colors, arranged in a simple repeating pattern. But in one particular stall, the pattern was violated: two tiles are the wrong colors, in a position right where you are staring if you go to do your business. Someone must have replaced some cracked tiles a few years back, and just not paid attention to the pattern. I'm not quite sure why this particular violation of symmetry bothers me; maybe it's just the irony that a company can create massive engineering feats like modern microprocessors, but not keep its bathroom tiling consistent.

This got me thinking about tilings in general. Nearly every real-life use of tiles is periodic, with translational symmetry: it consists of a pattern of tiles that is repeated over and over, so you can take a small section, shift it a certain distance, and you will see exactly the same pattern. But do tilings always have to be that way? What other options are there? Of course, you can just take an identical set of tiles whose angles evenly divide into 360 degrees, like squares or hexagons, and randomly color them to create an essentially random set of colorings on top of a repeating tile pattern. But I'm asking for a truly different concept: can we take a simple set of tile shapes and cover an infinite floor without ever repeating the pattern? Think of it like the graphical counterpart to the decimal expansion of π, 3.14159...: it is made of the ten basic digits, but never settles into a repeating pattern. In other words, do sets of tiles exist that can cover the plane aperiodically, without exhibiting translational symmetry?

Surprisingly, the answer is yes. Such tilings were first discovered in 1964 by Harvard mathematician Robert Berger, who found a set of 104 tiles that can only cover the plane aperiodically, without translational symmetry. But the most famous tilings in this class were developed in the 1970s by Roger Penrose, who found numerous examples of as few as two tiles which could cover the plane aperiodically. One famous example is based on trying to use pentagon-shaped tiles. You can't cover a plane with such tiles, of course, since the angles of a pentagon are each 108 degrees, and this does not divide evenly into 360, so a set of pentagons cannot intersect at a point without leaving gaps. But Penrose defined a small set of tiles that can be used to fill these gaps, consisting of a star, a diamond, and a small odd-shaped tile that looks like a boat. By combining these with pentagon tiles, you can fill the plane with a pattern that never displays translational symmetry.

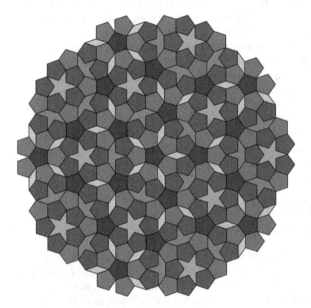

Figure 3-2. Penrose tiling[2]

Penrose tilings also have some other surprising properties: if you take any finite region, it will be repeated in an infinite number of other places, though not in any regular pattern. And some of these Penrose tilings also have rotational symmetry, which means that if you rotate the tiling at a certain angle, you will repeat the same tiling pattern. You can find lots of great illustrations of Penrose tilings at the *Wikipedia* page on this topic.

Initially, when I was researching this podcast, I thought the comparison to bathroom tilings would be an amusing joke, since nobody would seriously put in the effort to actually place an aperiodic tiling in a real-life building. But when reading about the topic online, I was surprised to see that in the 13th–15th centuries in the Islamic world,

---

[2]Sourced from Wikimedia Commons at https://commons.wikimedia.org/wiki/File:Penrose_Tiling_(P1).svg. Released into public domain by owner, user Inductiveload.

some architects actually used a set of five tiles of varying shapes known as "Girih tiles" that really could form an aperiodic tiling of the plane if placed properly. They consist of a decagon, an elongated hexagon, a bow-tie shape, a rhombus, and a pentagon. Almost every building known to use the tiles actually used them in periodic patterns, which are also possible, but at least one building, the Darb-i-Imam funerary complex in Iran, really does use an aperiodic pattern that can be extended into a full tiling of the plane. Did someone stumble on this by trial-and-error, or did early Islamic mathematicians actually achieve geometric breakthroughs that were lost until the late 20th century? We'll never know for sure.

Bees Vs Mathematicians
From Math Mutation podcast 152

Getting back to more realistic bathroom tilings, when I was younger, I spent a lot of time thinking about why we see hexagon-shaped tiles used so often. After staring at them for a while, you can probably figure out without too much trouble that it's all in the angles. The total angles around any point where tiles meet has to be 360 degrees. So if you want to tile using regular polygons, each one needs to have angles that divide into 360. You may recall that the formula for the total of the angles of a convex polygon is $180(n\text{-}2)$, where n is the number of sides. So this means that if you are tiling with regular polygons, triangles, squares, and hexagons are your only option, due to their 60, 90, and 120 degree angles. And let's face it, regular tilings of both triangles and squares are pretty boring to look at, with lots of straight lines forming regular matrices. So hexagons are a clear winner there.

Incidentally, this got me wondering, why do bees build hives out of hexagons? I bet it's probably easier for nature to evolve a regular pattern than to evolve something more complex, but there must be more to it. There's another important property of polygons we need to consider here: how can you enclose the highest possible area while minimizing the perimeter of a figure? The best answer is to use a circle, but that's cheating, since we were asking the question about polygons. (You might argue that it's not really cheating, since a circle is just a polygon with an infinite number of sides, but that's getting into some issues of philosophy.) If you have to use a finite regular polygon, you enclose more area the more sides you have, as that gets you closer to the circumscribed circle. And if you want to fill a plane without wasting any space, you need to choose a shape that allows a regular tiling – leaving hexagons as the best choice.

We're not done though. Can you spot the hole in this logic? Think for a minute about hidden assumptions. Why should bees be required to use a single shape, and why should they be required to make it out of straight lines? The idea that hexagons are still the best answer for efficiently dividing the plane into equally sized regions while minimizing the length of the lines used, even if irregular shapes and curves are still allowed, is known as the "honeycomb conjecture". It's actually been around since ancient times; Pappas of Alexandria speculated on this question in the 4th century AD. Surprisingly, it was not proven until the 1990s, by Thomas Hales of the University of Michigan. Hales showed that indeed, any partition of the plane into regions of equal area would have a perimeter at least equal to that of a regular hexagonal grid.

We also shouldn't forget that the hexagon shape of beehives is a bit of a simplification: like most of real life, beehives are actually a 3-dimensional structure. Every pair of hexagonal layers is offset from each other a bit, and the bottoms of the hexagon-sided cells are kind of pointy, made of three diamond-shaped panels. With these shapes, the layers are able to interlock; in an article on *Science News*, Ivars Peterson points out that this is kind of like the bottoms of two egg cartons stuck together. And this leads to another surprise: Hungarian mathematician László Fejes Tóth proved in 1964 that if they made the tips out of two hexagons and two squares, they would be able to enclose more volume in relation to the wax used. So here's a case where evolution didn't seem to get bees the right answer. Could that really be right? Well, the Tóth structure is relatively complex, so it's probably hard for nature to come up with the right one through random mutation.

But not so fast: we can't forget that mathematical results are in a kind of ideal space, not accounting for physical details, perhaps including the fact that cell walls have a nonzero thickness. To confirm Tóth's result, scientists Dennis Weaire and Robert Phelan did some experiments with detergent bubbles, which naturally minimize their surface in relation to the enclosed area, by pumping bubbles between two glass plates. When the walls were very thin, they indeed confirmed the result, forming beehive-like hexagon layers connected by tips using Tóth's hexagon/square structure. But when the liquid walls reached a certain thickness, it instead formed the three-rhombus structure seen in actual beehives. Thus, it's likely that when wall thickness is taken into account, Tóth's model was oversimplified. Proving once again that no matter how smart mathematicians are, nature is sometimes smarter.

A Brush with Evil
From Math Mutation podcast 102

Today I thought it would be fun to discuss pentagrams. You know the figure I'm talking about: basically you draw a pentagon, then draw all the diagonals connecting the vertices, and you have a pretty star shape. The pentagram seems to have had deep symbolic meaning in many cultures since ancient times, and most popularly brings demonic connotations to mind in our modern culture. Why do we find this shape so fascinating?

To start with, I think it's the easiest "interesting" picture to draw. Think about it: one lazy way to doodle, especially for the non-artistically-inclined, is to draw a bunch of dots and connect them. If you draw 2 or 3 dots, you get a simple line or triangle. If you draw 4 dots and connect them all, you get a quadrilateral with an X in it, which is not very exciting. But draw 5 dots with even spacing, connect them – and you have a pentagram! If the spacing is not even, you get some kind of distorted pentagram, except in the degenerate cases where several of your dots coincide or fall on a line. If you're aiming for a star shape, the pentagram star is also an easy figure to draw, in that you can draw the complete star in five lines without lifting your pencil, as I'm sure you learned in grade school.

Each intersection between diagonals of a pentagram divides the line into segments measurable by the "golden ratio", the famed mathematical constant φ equal to the ratio a/b such that $a/b = (a+b)/a$, which is $(1+\sqrt{5})/2$. You can also see that the center lines of the pentagram together form a smaller pentagon, in which you can draw another pentagram. In fact, if you look carefully, you'll see that in the 'arms' of the star, you can also inscribe a diminishing infinite series of smaller pentagons and pentagrams. If the starting pentagram was inscribed in a circle of radius 1, or had a circumradius of 1, then the nth iteration of smaller pentagrams has a circumradius of φ^{-n}.

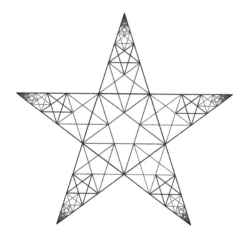

Figure 3-3. *Inscribed pentagrams and pentagons*[3]

Aside from these properties, the close association of the pentagram with the number 5 is probably another reason why human cultures have continuously found it fascinating. As early as 4000 BC it is said to have represented the "sacred feminine". To the ancient Hebrews, it represented the five books of the Torah. Perhaps because the 5 points can be put in correspondence to 2 arms, 2 legs, and a head, the top point of an upward-pointing pentagram can be said to represent the Spirit. According to the *Skeptic's Dictionary*, an inverted pentagram may be considered especially evil because it relegates the spirit to the bottom of the metaphysical heap. In ancient Greece, the mathematical cult of Pythagoras used pentagrams as a symbol of health, and wore them so followers could recognize each other.

The pentagram was a Christian symbol to the Roman emperor Constantine I, representing the Five Wounds of Christ. The knight Gawain in medieval literature inherited this use. Due to the ability to draw the full figure without lifting your pencil, the pentagram also became known as the Endless Knot, a symbol of Truth and protection against demons. A pentagram was used to contain the demon Mephistopholes in Goethe's *Faust*, which probably helped lead to its adoption by Satanists in more recent times. As a result, many Christians consider the pentagram in general to be demonic, regardless of which way it's pointing. To Wiccans and Japanese, however, the pentagram has no evil connotations, and merely symbolizes the five mystical elements.

Perhaps the most amusing comment I found in articles online is the most obvious. As you've probably seen in thousands of cartoons and children's drawings, the five-pointed pentagram star seems to be a near-universal symbol for the stars in the sky. But as the *Skeptic's Dictionary* points out, no star ever looks anything like that, whether to the naked eye or through a telescope. Strange, isn't it?

---

[3]Image used with permission under the Creative Commons Attribution-Share Alike 3.0 Unported license; original found here: https://commons.wikimedia.org/wiki/File:Embedded_pentagrams_(thick,_transparent).png. Copyright user KovacsUR.

Twistier Than You Thought
From Math Mutation podcast 168

If you're geeky enough to be reading this book, you probably know the basics of the Möbius Strip: take a long strip of paper, and tape one short end to the other, after rotating it 180 degrees. This results in an object that has only one side and one edge. If you start drawing a line in the middle, you will find that by the time it has returned to its starting point, your line traverses both sides of the paper. And if you start highlighting an edge, you will similarly find that all edges are covered by the time you get back to your starting point.

The Möbius Strip is named after August Ferdinand Möbius, a 19th century German mathematician and astronomer who had studied under Gauss. He was seen as quite talented in his youth, and became chair of astronomy at the University of Leipzig in 1816. My favorite anecdote about him is the fact that, despite being a department chair, he wasn't promoted to full professor very quickly, because he was not a good enough lecturer to attract students to pay to listen to him! If only modern universities followed similar policies, instead of torturing undergraduates with professors promoted only based on publications. Möbius ended up making major contributions to astronomy, analytical geometry, and topology. Ironically, his work on the strip, which was to make him a household name, was only published posthumously in a memoir, and the strip was discovered earlier (but apparently not described as completely) by another mathematician named Johann Listing.

Möbius strips are fun to play with. For a few surprises, try cutting one in half along its center, or cutting a parallel line 1/3 of the way from the edge. You will probably find the results don't quite match your intuition – you will end up with one long non-Möbius loop in the first case, and a linked Möbius and non-Möbius loop in the second. On the web you can find pages describing more general formulas for what you get from parallel cuts in a Möbius strip. A stranger experiment with a Möbius strip is to use it as tape for one of those punch-tape-based music boxes: the result will be playing a song the right way up once, then playing the same song with all the notes transposed to their opposites. One *YouTube* user has posted a video of this trick.

But aside from being an interesting curiosity, you might be surprised to learn that Möbius strips really do have practical applications. This isn't so surprising, actually, after you think about it for a minute: many elements of modern technology from the past century have used long tapes or belts of one type or another, and applying a Möbius twist allows both sides of an endlessly rotating strip to be effectively used without needing to remove and turn over the tape. In the case of a typewriter ribbon, it allows ink on both sides to be used up before the ribbon is worn out. In the case of a magnetic computer or audio tape, again it can allow recording on both sides. Those last two might not make much sense to you if you're not an elderly 46-year-old like me, but you can read about them in the history books. However, factories still need to use conveyor belts, until we all upload our consciousness into computers and cease to interact with physical objects altogether. In the case of a conveyor belt, a Möbius design allows us to ensure equal wear and tear on both sides, doubling its useful life. Various sanding and abrasive belts have similar issues. In a slightly more esoteric application, some music theorists have also found that when they try to describe chord relationships on a two-dimensional surface, the result is a structure that can be directly mapped to a Möbius strip.

And it's not just humans who have figured out how to make use of Möbius strips – they actually appear in nature. Analyzing an African folk medicine that seemed helpful for pregnant women, biologists discovered a key component was a protein called Kalata B1. Kalata B1 is a cyclic protein, consisting of 29 amino acids connected in a loop – and the loop has a single half-twist, resulting in a Möbius-like structure. Apparently this makes the protein especially stable, able to survive boiling temperatures, and makes it very complex for enzymes to interact with. Experiments have shown that this protein also has antimicrobial and insecticidal properties. It is suspected that these kinds of structures will have many pharmaceutical and agricultural applications.

Finally, we shouldn't neglect the uses of Möbius strips in the arts. You can see them in many paintings, most famously those of M.C. Escher, and of course a drawing of the Möbius strip is the basis of the green-triangle recycling symbol seen in many politically correct locations around the world. Whimsical stories about Möbius strips have also been a popular theme in science fiction short stories: one website on such things counts 31 well-known stories using the topic, including ones by luminaries such as Arthur C. Clarke, John Barth, Martin Gardner, and Lewis Carroll. I'll conclude this section with a more obscure work, an Internet poem by someone whose name (or pseudonym) is Eleanor Ninestein:

> The Topologist's child was hyper
>
> 'Til she wore a Möbius diaper.
>
> The mess on the inside
>
> Was thus on the outside
>
> And it was easy for someone to wipe her.

Squash Those Dice
From Math Mutation podcast 18

Recently I was playing in my weekly game of *Dungeons and Dragons* ("D&D"), and started thinking about the dice we use in that game. The main dice in D&D correspond to the five regular polyhedra, also known as the platonic solids. These are solid shapes that have the nice property that every side is a regular polygon, and every vertex is congruent to every other vertex. They consist of the tetrahedron, with 4 triangular faces; the cube, with 6 square faces; the octahedron, with 8 triangular faces; the dodecahedron, with 12 pentagonal faces; and the icosahedron, with 20 triangular faces.

Figure 3-4. *The five regular polyhedral dice*

What has always seemed odd to me is the fact that there are exactly five of these regular polyhedra. Why shouldn't there be any others? Why isn't some gigantic wacky 8,763-sided shape with nonagon faces, that nobody has bothered building yet, lurking out there? That die would certainly add some spice to our D&D games.

Years ago in a math class, I learned a standard algebraic proof that the five platonic solids were the only ones, involving equations describing the relations between the number of vertices, edges, and faces. But I never really had any intuition about why this was so. Thus I was happy to see, when skimming the web for more information on this topic, a simple geometric argument for why these five solids are unique.

To understand this argument, let's look at one of the vertices (corners) of one of these solids, a point where three or more of the polygonal faces meet. Imagine that you are frustrated with your latest die roll, so you take your palm and smash this vertex into the ground, and the faces break apart from each other, but still remain attached to the vertex point you started with. Look at the point on the ground where that vertex has landed. Around it are all the adjacent faces, with one angle from each in a circle around the vertex. Since these angles together, flattened in the plane, all fit around a single point, they must total less than 360 degrees. Strictly less than, and not equal, since at exactly 360 degrees they would have had to be flat in the plane to start with, and could never have been "folded" upwards into a polygonal shape. This is the key observation – the uniqueness of the five solids derives directly from this constraint, that the angles touching a vertex must total less than 360 degrees.

Why is that? Well, remember that at least three faces must touch to make a vertex, and we have defined regular polyhedra to require that all faces are identical. So now we know that each face must have corners with an angle strictly less than 120 degrees. But each angle of a hexagon is 120 degrees, and regular polygons have wider angles the more sides you add, so that means that each face of a regular polyhedron must be a pentagon, square, or triangle. In the case of pentagons or squares, you can only fit three around a vertex without hitting the 360 degree limit, so there is only one platonic solid with pentagonal faces, the dodecahedron, and one with square faces, the cube. In the case of triangular faces, each angle is only 60 degrees, so we can have three, four, or five of them at a vertex, defining our last three solids: the tetrahedron, octahedron, and icosahedron. The fewer we have at a vertex, the sharper the angle is when we "fold" them up to make a solid.

Now, those of you who are hardcore gamers may be confused by the fact that you have seen other dice in stores, such as the 10-sider, 30-sider, or 100-sider. But none of these are regular polyhedra: either not all angles are congruent, or the sides are not regular polygons. As for why we care about that restriction – well, that makes the math more interesting. Fair enough? Due to our insight about the angles around each vertex, we know that as long as we want our dice to be regular polyhedra, the basic five are the only ones we will ever get.

Crazy Dice
From Math Mutation podcast 155

In the previous section, we discussed regular polyhedral dice. These are formed from the five convex solids in which every side is a regular polygon, and every vertex is identical. There are the 4-sided tetrahedron, the 6-sided cube, the 8-sided octahedron, the 12-sided dodecahedron, and the 20-sided icosahedron. Why are these dice preferred for gaming, over more irregular shapes? You can see why these would be naturally fair dice, as their multiple symmetries make all the faces equivalent in some sense. But do we need such strict criteria to create fair dice, or dice with equal probabilities of landing on each side? If you think about it a little, the requirement of a Platonic solid is really too strict. A simpler criterion for a fair die is that it be *isohedral*: this means it is a convex polygon with all faces identical, and with each face having the same relationship with the shape's center of gravity. In other words, we can loosen the requirements that all faces are regular polygons, and that all vertices are identical.

So... what does that buy us? Well, quite a bit in fact. On the web you can find a nice site by Ed Pegg, who shows cool 3-D plots of dice with 24, 30, 48, 60, or 120 sides. But even more useful, he also discusses infinite families of isohedra with $2n$ sides, for any n value of 3 or more. The simplest ones are easy to visualize: just take a cone, but instead of a circular base, make the base a regular n-gon. Then put two of those cones base-to-base with the corners touching, and voila, you have a fair die with $2n$ sides. And note that you can easily use a $2n$-sided die to fairly generate random numbers from 1 to n, simply by numbering it up to n twice instead of uniquely numbering all sides. So no matter what crazy role-playing game you invent, you can include appropriate dice to generate random numbers in any range.

But as with any basic mathematical principle, there are always wise guys who have to find a way to violate it. So they make dice with funny numbers of sides like 5 or 7, but instead of using the simple formula we just mentioned, they find other tricks. For example, here's one way to make a 7-sided die that is actually manufactured & sold to gamers: visualize a cylinder standing upright like a soda can, but for the top and bottom bases, use pentagons rather than circles. Think about this for a minute – it means two of the sides, the bases, will be pentagons, but the other 5 will be rectangles. Can this really be a fair die? If you're very careful about the area of each of the sides and the distribution of weight within the die, you can probably even things out to make this fair. It's also unusual in that if it lands on the top or bottom, you can read the value on the face like a normal die, but if it lands on a rectangular side, what's facing up will be an edge, so those numbers will have to be labeled on the edges.

It's a bit trickier than you might think at first, though, to make an odd-shaped die fair. A naive calculation would just be to compare the area of the sides, and assume that if they are all equal, we're fine. You can quickly disprove this with a thought experiment: Imagine that in the 7-sided die we just discussed, the pentagons on the ends are cut at an angle rather than straight, so if you let the die rest on one, it will be leaning to one side and topple over. The probably of rolling on one of these pentagons effectively drops to 0, regardless of their area. In these odd shapes, you really need to account for the energy needed to topple the die over to the next side, as well as the likely momentum from a roll, to figure out not only if the die will land on that side, but also if it will stay there. Pegg calculates what he calls the Energy State Model to formalize this, which you can see in detail on his website.

In another article, Ivars Peterson points out one more wrinkle that affects the fairness of dice. In most nicer dice, the numbers are indicated by small dots that are drilled into each face. But the tiny amount of material removed in this drilling means that the weights of the faces are not perfectly symmetrical, and the side with only one dot is heavier than the side with six. So the six is more likely to be face-up than the one, and your gnome is a likely to slay that dragon a tiny bit faster. Is this effect really significant? According to numbers posted on the web, out of 10000 rolls, you can expect 1679 sixes, but only 1654 ones. Probably not enough to tip a casual game, but it is enough to concern the casinos: they actually fill in the dots with paint that has a weight precisely matching the missing plastic, to eliminate the issue.

So, while it takes a little work, you can make fair dice that generate just about any number. Maybe one day you will be clever enough to add yet another to the hundreds of types of dice registered at the U.S. patent office.

Wheels That Aren't Round
From Math Mutation podcast 209

Most of us are generally familiar with the concept that wheels are usually round. But do they have to be this way? What are the properties that make a round wheel useful? Yes, you might think that a decade of math podcasting has finally driven me insane, to question something so obvious. But math geeks are famous for requiring proofs of the obvious – and this is a case where common instincts might lead us astray. Now of course, for a wide variety of reasons, circular shapes do tend to make the best wheels. In certain cases, though, there is a more general class of figures that can be substituted for the circular shape, with some important real-life applications. These are known as *curves of constant width*.

To simplify the discussion and avoid the complications of axles, let's discuss simple rollers. Suppose you want to smoothly roll a large plank across the top of a bunch of logs. If the logs have a circular cross section, it's pretty obvious that the plank can roll smoothly along, without wobbling up and down. But what is the property that enables this? The reason for the plank's smooth rolling is that the circle is a curve of constant width. This means that if you put parallel lines above and below a circle and touching it, the distance will be a constant, the diameter of a circle. However, a surprising fact discovered by Euler in the 18th century is that there are many other curves of constant width that could be used instead and still allow smooth rolling.

The most famous non-circular curve in this class is known as the Reuleaux Triangle, a kind of equilateral triangle with rounded edges. To create one, start with an ordinary equilateral triangle. Then, for each vertex, replace the opposite side with the arc of a circle whose center is that vertex, and whose radius matches the side of the triangle.

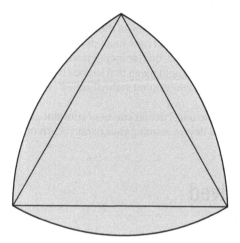

Figure 3-5. Reuleaux triangle

If you think about it for a minute, you should see that this curve will be of constant diameter: if a plank is rolling over the top, at any given moment either the plank or the ground will be touching a vertex, and the opposite surface will be touching a curved edge. Since the circle used to form that curved edge is defined as the set of points equidistant from its center, the opposite vertex in this case, the distance between the plank and the ground will be a constant value equal to the circle's radius. Thus, logs with a Reuleaux cross section will be rolled over just as smoothly as circular ones.

As you can probably see from how we constructed it, the Reuleaux Triangle is just one representative of a large class of curves of constant width. Take any regular polygon with an odd number of sides, and replace the side opposite each vertex with an arc of a circle centered at that vertex. There are also many other curves in this class, with more complicated construction methods; you can read up on these on the web if you're curious.

The surprising discovery of this large class of shapes has led to some useful real-life applications. Reuleaux, the 19th century engineer for whom the triangle is named (despite Euler's earlier knowledge of it), became famous for investigating a variety of uses based on converting circular into other types of motion. Later this led to applications in mechanisms as diverse as film projectors and automotive engines. Since a rotating Reuleaux triangle traces a shape that is nearly square, it has also been used to construct a special drill bit that enables woodworkers to drill square holes. By basing the drill on other curves of constant width, a similar method can be used to drill pentagon, hexagon, or octagon-shaped holes as well. This shape has also been used in the design of pencils, with the claim that the constant diameter but non-circular shape provide a comfortable grip while reducing the chance of spontaneously rolling off a table. And in several countries, non-circular curves of constant width have been used as the shape of coins, with their constant diameter providing advantages in the design of vending machines.

But one of the most useful aspects of the Reuleaux Triangle and related shapes is as a non-circular counterexample, forcing us to question basic assumptions about simple geometric properties. According to some sources, engineers working on the doomed space shuttle *Challenger* tried to verify the cylindrical shape of some components by measuring their width at various sampling points, not being aware of the existence of non-circular curves of constant diameter. Too bad they didn't have math podcasts back then, though technically the engineers could have read Martin Gardner's classic essay on the topic. Anyway, if the shapes were not circular, this would mean that various types of stress would affect the parts unevenly. This may have contributed to the shuttle's eventual destruction.

So, be sure to think about the existence of these non-circular curves of constant width, next time you are assembling a mechanical device, minting your nation's currency, or designing a space shuttle.

The Future That Never Arrived
From Math Mutation podcast 73

If you've gone to Disney's Epcot center, or seen other visions of the future created in the 1950s and 1960s, you've probably seen a geodesic dome. These are the big domes, made up of triangles that were thought by many half a century ago to represent the future of architecture. They still do look cool, but never quite caught on, and most of us still live in rectilinear houses. What are these geodesic domes, though, and what makes them so interesting?

In general a dome is an efficient design, as judged by the relationship of interior volume to surface area, since in three dimensions, spherical figures enclose the most volume with the least surface. A geodesic dome is built using a network of linear elements forming portions of great circles around a sphere, where a "great circle" is a circle like the equator that has a diameter equal to the sphere itself. The intersecting network of great circles forms a large number of triangular elements. The most important property of these geodesic domes is that these great circles distribute stress across the structure, making them naturally very strong. However, due to the large network of intersecting elements, small flaws can cause stress to be transferred in very odd ways: *Wikipedia* references an incident where a snow plow bumped into one of these buildings in Princeton, New Jersey, and the damage appeared on the opposite side. But properly built domes have been seen to stand up extremely well against earthquakes, hurricanes, and other risks. Since this self-reinforcing structure is inherently strong, if you do it right, it is apparently possible to build one out of very lightweight materials; I spotted one website that shows you how to build one out of straws and gumballs.

But you've most likely heard about geodesic domes more recently due to 'buckyballs,' the 60-atom carbon molecules discovered in the last few decades that have geodesic-dome-like shapes. Their extreme structural stability leads to many potentially interesting applications, such as using them to contain and transport simpler molecules, including densely packed hydrogen. Buckyballs were named after architect Buckminster Fuller, thought to be the inventor of geodesic dome architecture. He certainly patented and popularized them, though there is some dispute about whether certain German designs from the 1920s technically beat him to it.

Fuller led a successful but very unusual life. While he gained international fame from his exotic but extremely strong dome-shaped buildings, he attributed more universal significance to his geometrical observations. These included principles of close-packing spheres, tetrahedrons, and octahedrons, which he called "tensegrity structures", in addition to understanding geodesic domes. He considered himself not just an architect, but a philosopher, and believed that understanding the interdependent nature of these geometric structures would lead naturally to environmentalism, universal government, and world peace. He called this philosophy Synergetics, explained as "The integration of geometry and philosophy in a single conceptual system providing a common language and accounting for both the physical and metaphysical."

I tried to read his long, ponderous book on the Synergetics, which describes this philosophy in detail, but didn't quite make it through. As *Wikipedia* states, Fuller often created "long run-on sentences and used unusual compound words (omniwell-informed, intertransformative, omni-interaccommodative, omniself-regenerative) as well as terms he himself coined." He also considered his own thoughts and actions so significant that he made an effort to document every 15 minutes of his entire life between 1915 and 1983, titling the resulting work the *Dymaxion Chronofiles*. These files included items ranging from personal journal entries to dry cleaning bills. But the fact that he went off the deep end philosophically should not detract from his architectural achievements.

So anyway, if these geodesic domes are so great, why aren't we all living in them in the 21st century? To start with, domes have some obvious downsides: the sloping shapes mean that you will have a lot of wasted areas without enough headroom to be practical living space. And most building materials come in rectangular shapes, so it's hard to build a dome without a lot of waste. Furthermore, laws and regulations regarding things like fire escapes, windows, and electrical wiring are ill-suited to these structures, adding extra expense. Similar issues come due to the geometric conflict with commonly constructed furniture: where can you put a sofa in a dome without wasting the space behind it? None of these problems were enough to stop Fuller himself from living in one of his domes, of course, but the rest of us who aren't quite as rich as him have to consider these practical issues.

CHAPTER 4

■ ■ ■

Deeper Dimensions

I still remember the first time I read Edwin Abbott's classic 19th-century novel *Flatland*, back when I was in the 6th grade. *Flatland* tells the story of A. Square, a square-shaped creature who lives in a world that exists entire on a two-dimensional plane: he can move forward, backward, left, or right, but has no idea that the directions of up and down even exist. All he can see is other objects in his plane, which all look like mere lines from his plane-confined point of view. Eventually he is visited by a sphere from the 3rd dimension, and is amazed by its "tricks" such as the ability to get smaller or larger, depending on which of its cross-sections is intersecting the Flatland plane. Eventually the square comes to accept that there really are dimensions beyond the two that he is used to.

As I reviewed my list of *Math Mutation* episodes in preparation for assembling this book, I was surprised to realize that I didn't actually have an episode directly focused on discussing *Flatland*. In the back of my mind, I probably figured that anyone interested enough in math to listen to my podcast must have already read that book. If you haven't, I strongly recommend that you do! My description in the paragraph above certainly fails to do it justice, though it should be sufficient for you to be able to comprehend the related discussions in this chapter. Note that *Flatland* is old enough to be out of copyright, so you can find the complete text for free online, in case buying this book has already expended your reading budget.

Reading *Flatland* was an eye-opening experience. Could it really be the case that we are just like the Flatlanders, unable to detect the real higher dimensions of the universe due to our misfortune of originating in a limited three-dimensional hyperplane? How many amazing discoveries are out there, perhaps mere inches away from us, that we are simply failing to notice? How would our lives be transformed if we could figure out a way to interact with these higher dimensions? Ever since then, I have eagerly absorbed all the books, articles, and stories I could on this mind-blowing topic. Here I share some of my favorite ideas and observations on the possible existence of higher dimensions, and the strange properties they would possess.

Making Flatland Real
From Math Mutation episodes 4, 8, 74, and 175

The original Flatland was just an infinite plane, where the inhabitants could travel around at will. While the novel was brilliant, its author, Edwin Abbott, didn't give much thought to two-dimensional energy sources, ecology, biological plausibility, or other mundane

© Erik Seligman 2016
E. Seligman, *Math Mutation Classics*, DOI 10.1007/978-1-4842-1892-1_4

considerations that make life possible in our universe. None of this was necessary to express his basic metaphor, using two-dimensional creatures ignorant of the 3rd dimension to question how we view our own world. But it still would have been nice for Flatland to be a little more realistic. In the years since Abbott's original publication, numerous other authors tried to extend Flatland into a slightly more 'realistic' model, if you can call it that. For example, it would be more analogous to our experience if the planar creatures lived on the rim of a disk in the universe, just like we live on the surface of our planet. Three subsequent authors created particularly notable works in this genre: C. H. Hinton, Dionys Burger, and A.K. Dewdney.

C.H. Hinton was actually a contemporary of Abbott, and wrote his works on the fourth dimension soon after *Flatland*. Less famous than Abbott, probably because he was not quite as talented a storyteller, he wrote about another two-dimensional world of his own invention. In Hinton's world of Astria, the creatures live on a circular planet on the surface of a large three-dimensional bubble. In many ways this predicted Einstein's later description of space-time, where our (perceived) 3-dimensional universe is curved in the fourth dimension. Hinton also made an effort to deal with the various difficulties of living in such a world. For example, he described how one could create wheeled vehicles. You can't have a wheel rotating around an axle, since a round wheel would have its movement impeded by any connected rod, so you would have to sit a cart upon some free-moving rollers. Someone in the back has to pick up the rollers as they emerge and pass them to the front again. Even worse, if you are trying to pull this cart forward with a rope and some kind of 2-D horse, you need to unhook the rope every time you want to place a new roller down. Think about it for a minute – a single line of rope in two-dimensions is as much of a barrier as a solid wall of rope would be to us!

While he was very realistic in some ways, Hinton also described the intelligent beings in his world as geometric shapes and used them as a vehicle for social commentary, much like Abbott did. Unlike Abbott's male polygons and female lines, in Hinton's world, all people are right triangles, with one leg parallel to the ground and the other perpendicular to it. This means that everyone has a straight side, where they can communicate with others, and a pointy side, which they use for self-defense. Men have their pointy side facing left, and women have their pointy side facing right. Thus, in a heartwarming illustration of family values, men and women naturally fit together: a man and his wife can stand side by side and defend each other from the world with their pointy ends. Two men or two women cannot join together in such a way, so I think this world might be ruled illegal by the Supreme Court of California. But this depiction also makes a not-so-subtle statement about women's rights issues of the day: while it may not be obvious to the inhabitants, it's clear to us higher-dimensional beings that although Astrian men and women may seem different, they are actually the same, and in the third dimension we could easily transform one into the other. So maybe California would approve after all.

Hinton was especially interested in the fourth dimension as a possible explanation for mystical or psychic phenomena. If you can travel or exert force in the fourth dimension, all sorts of actions would be possible that seem to violate the laws of our three-dimensional world. In one of his later Astria stories, the inhabitants realize that their planet's orbit is slowly degrading, and their disk will eventually fall into their sun. To save themselves, the Astrians realize that their bodies extend into the third dimension, where what they call their 'soul' lies. By concentrating hard, they can invoke their natural psychic-like powers and cause some motion in that dimension. By spreading a religious awakening about this mysterious power in their souls, and having all the enlightened people work together, they manage to use their three-dimensional powers to save their planet.

A few generations later, in the 1960s, Dionys Burger wrote his own unofficial sequel to *Flatland*, titled *Sphereland*. *Sphereland* attempts to expand this story into an allegory about curved Einsteinian spacetime, and introduce readers to the possibility that the fourth dimension not only exists, but that our space is curved in that direction, as physicists now believe is the reality. It is revealed that rather than living in a generic plane, the Flatlanders live along the rim of a large circular planet, and there are other distant circular planets floating about in their two-dimensional space. There is a very amusing discussion of Flatland's Age of Exploration, when various expeditions discover properties of distant lands, and finally when explorers heading east and west unexpectedly encounter each other, prove that their world is a disc. In "modern" times in Flatland, an astronomer starts measuring large triangles, and discovers that some of them have angle sums greater than 180 degrees. He is thought to be a sloppy or crazy scientist by his fellow Flatlanders, since everyone knows that geometry has proven a triangle's angles always total 180 degrees. Convinced that his measurements are accurate, the scientist works with a hexagon, the grandson of *Flatland's* square, and together they discover the truth, that their universe is curved in the third dimension. Just as our universe is curved in 4-dimensional space, Flatland is curved in 3-dimensional space, and the Flatland universe sits on the surface of a sphere.

Probably the most realistic attempt to describe a two-dimensional world is A.K. Dewdney's *The Planiverse*, which was published in 1984. In addition to continuing with the concept of a "realistic" disc-shaped planet floating in a universe of similar planets and stars, Dewdney dismissed with the rather silly concept of living beings as polygonal figures, and tried to come up with plausible two-dimensional animals. He wrestled with numerous basic issues that would result: for example, think for a moment about your digestive or circulatory systems. These are basically a bunch of tubes running through your body. But in two dimensions, any kind of tube, intestine, or blood vessel essentially splits a shape in half. How could living beings exist? Dewdney describes a clever solution for this: the zipper organ. Like a zipper, two sets of cells can come together in groups while others nearby pull apart, thereby holding the whole body together as blood or food is being passed along. At any given moment, a large subset of the cells is connected, and the creature is held together.

Another example of an obvious but tricky question addressed by Dewdney is: how can someone build a house in a 2-D world? It's impossible to walk past a structure built on the surface of this circle-world, without climbing over it. I would not like having every transient in Portland climbing over my roof at all hours. I'm not sure people would appreciate having to climb over every building in town to go for a walk either. Dewdney's solution is that houses are built into holes in the ground. You always have to keep the door open though, except for the short time when someone needs to cross over – otherwise, it forms a perfect seal, keeping out the air. However, think about what happens to this two-dimensional burrow-house when it rains – you have to close the door, and it has to be 100% sealed, or else your home will turn into a swimming pool. Thus, your house had better hold enough air for you to survive until the end of a rainstorm.

So, if you enjoyed *Flatland* but really want to explore the inherent complexities and contradictions of a two-dimensional world, Hinton, Burger, and Dewdney are a great place to start. Hinton and Burger go a little further than Abbott did, though they are still primarily focused on philosophy and metaphor. Dewdney provides hardcore science-fiction realism, if that's something you prefer. Either way, if you enjoyed *Flatland*, all three of these additional authors deserve a place on your bookshelf.

Is Flatland Doomed?
From Math Mutation podcast 53

Now let's dive a little deeper into the "realistic" physics of a two-dimensional world, as explored in A.K. Dewdney's Planiverse. What would be the effect of fewer dimensions on things like chemistry and physics? How would the laws we're familiar with be subtly warped by the lack of dimension?

One example Dewdney points out is the familiar inverse-square law that applies to the strength of forces like gravity over distance. You may recall from high-school physics class that in our three-dimensional universe, if you multiply the distance between two objects by a factor of n, the gravity between them diminishes by a factor of n^2. Probably like me, you accepted this as a fact, used it in formulas, and just moved on. But is there a reason why it works like this? Why is it inverse-square, and not inverse-cube, or just inversely proportional? There is actually a reasonable explanation that is intimately tied with the dimensionality of our universe.

Think about a point of light on the inside of, say, a large-radius beach ball. The area it illuminates is the inside surface area of the beach ball, proportional to the square of the radius. Now imagine that the beach ball's radius expands by a factor of three. The inner surface area increases by a factor of nine, or three squared. But the point in the center hasn't changed – the same amount of light must now be spread across nine times the surface area. In other words, the inverse square law happens because as you move farther away, the same amount of energy must spread over an area that squares with the distance in order to produce the same effect.

So what does this mean in two dimensions? Well, our two-dimensional "beach ball" would be a circle, and the light would just have to illuminate the outer perimeter, whose length is directly proportional to the radius rather than its square. So rather than an inverse-square law, the two-dimensional universe has a simple-inverse law. This means that forces and energy diminish much less rapidly over distance in the two-dimensional universe. But does this simple-inverse law have more drastic consequences?

One fact critical in basic physics is that an integral, or continuous summation, from 1 to infinity of the inverse-square function converges: that is, values diminish so rapidly that the sum total of all the energy it would take to separate two objects to an infinite distance has a finite sum. This is why we have the concept of "escape velocity", a finite surface speed an object can have to take off of the earth's orbit. It also helps explain why a finite amount of energy can permanently sever an electron from its atom, and similar small-scale effects that facilitate chemical reactions.

But the integral from 1 to infinity of the simple inverse function does NOT converge. This means that in our two-dimensional universe, there is no escape velocity, and our 2-D NASA has a much harder time getting off the ground to explore space, though with enough energy we can reach any finite point. And it also means that chemical elements are much more stable. In fact, much of our chemistry and physics would be impossible, so our poor two-dimensional Planiversans would find themselves in a much less rich universe than ours, even after you take into account their lack of dimension. Actually, taking this information into account, we should seriously doubt whether our two-dimensional beings can exist at all, or their world is just too limited by this simple-inverse law.

However, all it not lost – let's reexamine our original reasoning. Could it be that somehow our 2-D universe will still obey an inverse square law? One possibility is that it's a 2-D subset of a truly 3-D universe: so our hypothetical point of light has to illuminate not just the circle around it, but a spherical area above and below the plane, that the inhabitants are not aware of. Or, alternatively, there is no reason to assume that 2-D space is empty: maybe there is some kind of space-filling aether, like people used to once hypothesize in our own universe that partially absorbs energy and causes it to more rapidly diminish. You can probably think of other ways our two-dimensional universe might rescue the inverse-square law, and still have lifelike chemistry and physics despite its lack of dimension.

In any case, this thought experiment shows that when changing the dimensionality of universes, Abbott's original idea of reasoning by analogy has a few flaws. We would need to think very carefully about changes to physics and chemistry when trying to realistically describe a two-dimensional universe.

Visitors from the Next Dimension
From Math Mutation podcast 17

What would it look like to us if a four-dimensional creature passed through our space? A fourth spatial dimension can be kind of hard to wrap your mind around, so the best way to answer these kinds of questions is often by analogy, thinking about what three-dimensional shapes would look like when passing through a two-dimensional world. Try to think about the situation of residents of Edwin Abbott's Flatland. They can travel in their ordinary two dimensions and observe things happening on their plane, but have no way to detect or envision what is beyond that. If any three-dimensional creature or object passes through the plane of Flatland, they can see cross-sections where it intersects their plane, but cannot see anything more.

So let's think about what a three-dimensional sphere would look like when passing through a two-dimensional world. When it first touches the plane, it starts out as a point, to the eyes of the Flatlanders. Then, as the sphere travels through the plane, it becomes a circle, getting bigger and bigger until it is halfway through. Then it starts shrinking again, until it becomes a point and disappears. By analogy, we can then see what will happen when a four-dimensional hypersphere passes through our space. It will start out as a point, which expands into a sphere, from our point of view. The sphere will gradually grow until it is halfway through our space, then start shrinking again, until it finally disappears.

What's more fun is to speculate about less regular higher-dimensional shapes passing through our world. Suppose a four-dimensional creature had some limb analogous to a human hand. What would this look like if he stuck it in our space? Well, let's go back to our two-dimensional example for an analogy. Suppose you stuck your hand into the plane of Flatland. What would it look like to the natives? At first, it would start out as five distorted circles, as your fingers intersect the plane. The circles would get slightly fatter and then, suddenly, when the main part of your hand reaches the plane, they would flow and merge together into one large blob. So, if the four-dimensional creature stuck its hand into our space, we would start by seeing a number of small irregular spheres. They would get slightly bigger, and then suddenly merge together into a large blob, as the four-dimensional 'palm' reaches our space.

Speculations like this are what lead many New Age types to think that UFOs, ghost sightings, or angel visitations might be higher-dimensional creatures passing through our world. Personally, I kind of doubt it – especially since most of these "sightings" are of humanoids, not of bizarrely odd-shaped beings that split and merge like we described in our hand metaphor above. But I suppose anything is possible.

Will the Real Fourth Dimension Please Stand Up?
From Math Mutation podcast 30

As we have been discussing, the fourth dimension is a new direction that is somehow perpendicular to length, width, and height, our usual X/Y/Z coordinates in Cartesian space, and thus represents a totally new measurement or direction of travel that we cannot perceive in ordinary life. But often we will hear people speak of Time as the fourth dimension, and indeed, the concept of time does meet our definition of a dimension in some sense. So, should we think of time as just another dimension, like our X, Y, and Z, or is it a different concept?

I should start by clarifying that I'm not talking about strict mathematical definitions here. In some sense, just about any number you can assign to some aspect of an object can be considered a dimension: color, brightness, voltage, etc. You will often hear of computer scientists thinking of modeling some set of parameters in an "n-dimensional" space, meaning that they are optimizing the set of numbers that describe some object, perhaps trying to come up with the fastest, cheapest, etc. Also, modern string theorists have hypothesized that the universe needs 10 or 11 dimensions. But in this podcast, I'm talking more of our innate sense of a "dimension", as some fundamental measurement that is in some sense as meaningful as length, width, and breadth. Does Time meet this criterion?

The most basic way to answer is to simply ask if objects are required to have a measurement of time, similar to their measurements in the other dimensions, in order to exist. If I tell you I have an object that is five feet long, and 0 feet wide, then it must be a figment of my imagination: something of 0 width cannot exist. Duration, the measurement of how much time an object exists for, does seem similar: if an object exists for 0 seconds, then it does not exist, any more than an object of 0 width. Everything must have a length, width, height, and duration.

Another intuitive criterion for labelling a dimension is: can Time be interchanged with one of the other spatial dimensions? For example, we can switch the X and Y coordinates of an object by turning it on its side. Can we do the same with Time and one of the spatial dimensions? If you think about it, the answer seems to be yes in some sense. Let's look at the metaphor of Flatland, a two-dimensional world we have mentioned in earlier podcasts. Flatland can be viewed as a large sheet of paper, where various two-dimensional creatures, such as living squares and triangles, can travel around, but have no idea there is another direction of travel that would let them leave the paper. If we watch the movie *Flatland: The Film*, we see the lives of these two-dimensional creatures unfold over the period of a couple of hours in time.

Now imagine that we like the movie so much that we capture each frame on our computer and print it out, sticking the frame printouts in a pile on our desks. Looking at the stack of printouts, we see that the top one is the first second of the movie, underneath it is the next second, and so on; when we get to the bottom of the stack, we are at the end of the movie. In other words, we have converted the time dimension of Flatland into the spatial dimension of height. If we imagine that the frames are printed on glass, we might be able to see the snakelike path of the starring square in this large solid glass cube on our desks. This path is known to physicists as a "world line", tracing out the activities of some solid creature through time in a static representation. There's actually a good illustration of a Flatlander's worldline (though they don't label it as that) in the Worldline article on *Wikipedia*. We three-dimensional creatures theoretically have worldlines too, though to actually depict them in a visual graph, we would need a fourth spatial dimension other than time.

However, we don't need to get nearly so exotic to describe worldlines. I do think Flatland is cool, so I try to shove it into an article whenever I can. But just think about tracing your favorite stock price in the Wall Street Journal. If you wait to read its value each day, you are essentially observing a one-dimensional quantity, the stock value, over time. On the other hand, if you look at a summary graph of the stock price over the last year, one of those crooked-line things that is all over the financial section, you are essentially seeing the stock's worldline, with the time dimension translated to the horizontal spatial dimension.

So, if worldlines do represent the transfer of a time dimension into another spatial dimension, is it an open-and-shut case that time really is just another dimension? I don't think so. However physicists try to sound sophisticated about how time is just another term in their equations, or an abstract measurement of the direction of progress of entropy under the laws of thermodynamics, I think we all agree that there is something fundamentally different about time. However we can transpose or represent it in a theoretical sense, it's the only dimension that has us all handcuffed in some cosmic prisoner transport, forcing us to continuously travel in one direction at about the same speed, regardless of what we want to be doing.

Ultimately, this kind of discussion starts to sound like the silly arguments over whether Pluto is a planet – it all depends how you define the words. In some ways, time clearly is like other dimensions, and viewing a worldline graph, it is hard to see how we could not consider time a regular dimension. On the other hand, the unique way our conscious minds experience time cannot be replaced by any other dimension, so in some sense it is a different concept. As for whether you want to think of time as a dimension or something apart from the dimensions – I don't think math or physics can ever come up with a definitive answer.

A Four-Dimensional House
From Math Mutation podcast 111

Recently I reread the classic 1940 Robert Heinlein story, "And He Built a Crooked House". This was a clever tale about an architect who decided that he wanted to build a four-dimensional house. Now at first you might object, saying that in our world we only can construct objects in three dimensions. If you're a little sneakier, you might point out that all houses are four-dimensional – if you regard time as the fourth dimension. As long as the house doesn't instantaneously disappear, it has a measurement in the time dimension.

But Heinlein's architect, Quintus Teal, wanted to build a true four-dimensional house, extending into a theoretical fourth spatial dimension that is truly perpendicular to our common x, y, and z axes, though currently imperceptible to us. To do this, he started by looking at the metaphor of trying to build a three-dimensional cube out of two-dimensional paper. I'm sure that you've seen an 'unfolded' cube at some point: on a piece of paper, you can draw six squares, in a shape like a cross. Then you can cut that six-square shape out of the paper, fold it up, and make it into a cube. So, reasoning by analogy, shouldn't you be able to do something similar with a four-dimensional hypercube? Just like you could draw the cross-like unfolded cube on a piece of paper, you should be able to build an unfolded hypercube in three-dimensional space. This would consist of eight cubes, in a three-dimensional version of that cross shape, essentially a four-cube-tall tower, with four side cubes jutting out of the second cube down from the top.

Figure 4-1. An unfolded hypercube

Now, just like the unfolded cube is still just a two-dimensional shape until you fold it in the third dimension, the unfolded hypercube is just an odd three-dimensional shape, until you fold it in the fourth dimension. So while Teal's concept for a house may have been architecturally interesting, it wasn't a true hypercube. On the other hand, you may recall that Heinlein was a science fiction author, so he had a way around that slight difficulty. The house happened to be built in earthquake-prone California. So one night, after a major earthquake, Teal came to the house and found that all he saw was a cube! In the night, his unfolded hypercube had folded in the fourth dimension. And just like a two-dimensional being would only see a plain square if there was a cube sitting on his plane, Teal could only see the "bottom" cube of the four-dimensional structure.

When Teal and his friends entered the house, it got even more confusing. As they travelled around the house, they sometimes were turned at odd angles in four-dimensional space, and re-entered the three-dimensional world at some other spot in 3-D space. You can imagine a two-dimensional flatlander reaching the edge of a cube: he can't even conceive of turning in the third dimension, so really has no control over whether he goes straight along the plane, or ends up creeping up the side of the cube. So he might think he is traveling the distance of three rooms, as he travels up the side, over the top, and down the other side – and then find himself on the opposite side of the room which he thought he originally entered, facing the front door, as he has arrived at the bottom square from the other direction.

There is at least one major flaw in the story though, even if you accept Heinlein's slightly odd reality. For the unfolded hypercube to get folded up into a real hypercube, it must mean that the earthquake was some kind of multi-dimensional super-earthquake, capable of exerting force in the direction of the fourth dimension. But if this were the case, then objects near the folded house would certainly have been subject to the same forces – so they should have seen lots of nearby buildings missing and holes in the landscape, places where objects were flung off our three-dimensional hyperplane into other parts of 4-D space. Looking at the metaphor of a 2-D world again: how likely would it be that an ordinary earthquake would take an unfolded cube you drew on a piece of paper and fold it up into a cube, without any other visible effects on the nearby paper?

In any case, I wouldn't recommend attempting a four-dimensional house in today's real estate market. You might be better off saving on building materials and downgrading to two dimensions instead.

Turning Around in Time
From Math Mutation podcast 63

I'm sure that if you are the kind of person who is reading this book, you have occasionally speculated about time travel. If time is truly another dimension, just like our ordinary three dimensions of space, then should we be able to change our direction? Are we really stuck traveling forward at a more-or-less constant rate, unless we can go fast enough to benefit from the distortions of relativity? It's fun to think about what it would look like if we could just decide to make a U-turn in time, like we can in space.

To make this example concrete, let's suppose I have a doctor's appointment at noon, and it takes me 10 minutes to walk down the street to the doctor's office. I look at my watch and notice that it's noon already. No problem, due to my time-walking ability. I step outside, spend 5 minutes walking halfway to the doctor's office, then make a U-turn in time, and walk 5 more minutes, traveling backward 5 minutes in time as I continue walking down the street toward the doctor's office in space. At noon, I am safely at the office, and start walking forward in time again, entering for my appointment.

What would this look like to a neighbor couple sitting on their porch, watching me walk down the street? From my house, they would see me emerging at noon, and arriving halfway down the street at 12:05. But from the doctor's office, they would see an odd sight. There would be a backward-me walking backward down the street toward my house, at the same time as the first me is walking forward. Think about it: at noon, my backward-traveling-self arrived at the office. At 12:01, I was one minute away from the office, traveling backward in time. At 12:02, I was two minutes away, and so on. But the most bizarre event happens at precisely 12:05, the backward-me and forward-me meet and merge, then suddenly disappear! At 12:06, I do not exist in the middle of the street, since at 12:05, I turned around in time. The neighbor couple will think I have suddenly disintegrated after a collision between my forward-self and backward-self. Eventually, of course, I will travel forward again and reach this point in time while in the doctor's waiting room. In fact, if it's a typical doctor, I will experience many future epochs of time in that waiting room, but that's another topic.

This kind of turning-around in time sounds absurd when described in terms of a person. But, strangely enough, many physicists believe they have observed precisely this phenomenon in the area of subatomic particles. A well-known type of interaction occurs when an electron and a positron collide, releasing a photon. And similarly, it is possible for a photon to spontaneously break down into an electron and a positron. According to a theory first proposed by the famous physicist Richard Feynman in 1949, we can also just view a positron as an electron traveling back in time. So when an electron and a positron collide to generate a photon, what is really happening is that the electron is hitting a photon and turning around in time, just like when I turned around in time in the middle of the street. And the positron it hit is just the same electron, travelling backward. Similarly, when a photon spontaneously splits into an electron and a positron, what's really happening is that a backward-electron hit the photon as it travelled backward in time from the future, bounced off it, and became an ordinary forward-moving electron.

So, even if I can't change direction in the fourth dimension to catch up when I'm late for an appointment, such direction-changing on the subatomic scale is a real phenomenon that has been observed, at least according to some theories.

11-Dimensional Spaghetti Monsters
From Math Mutation podcast 38

You have probably heard the term "string theory", the exotic theory dominating the study of fundamental particles in modern physics. It claims that rather than point-like basic particles, the fundamental constituents of matter are tiny vibrating loops of string. One of the many bizarre properties of this theory is the claim that these strings that make up our universe are 10- or 11-dimensional, but the dimensions other than the four spacetime ones we are familiar with are not detectable by our available methods. These extra dimensions are required to make the equations of string theory work out, although it sounds hard to believe. How can we have so many dimensions? How is it possible to be living in an 11-dimensional world, but think we only have four?

One easy way to look at this is by returning to our familiar metaphor of Flatland, an infinite two-dimensional plane inhabited by living circles, squares, and similar figures. Flatlanders are able to move about within their plane, but never able to leave it or even sense anything off of it, and thus have no concept of the directions 'up' and 'down' off the plane. But suppose that their 'plane' is actually the surface of a gigantic soap bubble floating in our space. In that case, the world of Flatland is really just the surface of a three-dimensional bubble, though the inhabitants are unaware of this fact. This is similar to the concept proposed by Dionys Burger in *Sphereland*. The Flatlanders are restricted to a two-dimensional subspace of the true universe. Similarly, it's possible that we are living within a four-dimensional subspace, a giant soap bubble floating in the true 11-dimensional universe.

Another explanation is that the extra dimensions are compact, rolled up into a space so small as to be undetectable. To better understand this, let's look at the Flatlanders' dimension-challenged cousins, the Linelanders. The world of Lineland is one dimensional – its inhabitants are points and small segments that can move back and forth on a single line, and are not even aware of a second dimension. On this line they can never pass each other, since there is no way to "get out of the way" of an approaching visitor. But suppose that Lineland is actually an infinitely long but very thin piece of spaghetti. This means it's actually three-dimensional, though it is infinitely large in the length dimension and very tiny in the other dimensions. The Linelanders are actually made up of little rings and tubes around the surface of the spaghetti, rather than true points. But being unable to detect any phenomena at the scale of a spaghetti-width, have no idea of the existence of these extra dimensions. We might be in a similar situation, except that our spaghetti is 11-dimensional.

So, are we wandering around the surface of a gigantic soap bubble in multidimensional spacetime? Are we foolishly traversing the four-dimensional surface of an 11-dimensional piece of pasta, never knowing of the vastly greater number of directions in space? Or are string theorists playing a silly game with their equations that has no relationship to reality? You will be sure to win a Nobel Prize if you can figure out and prove the real answer.

Your Five-Dimensional Kitchen
From Math Mutation podcast 133

In this chapter, we've been talking about the weird properties of multidimensional geometry, with the clear implication that these odd situations could only arise as a result of wild speculation or detailed calculations by advanced physicists. But did you know that you may very well have items in your kitchen that make use of five-dimensional geometry? Some types of coatings for nonstick cookware are made with a recently discovered and very useful material called quasicrystals.

To understand quasicrystals, let's start by reviewing the concept of an aperiodic tiling, which we described back in Chapter 3. A periodic tiling is a set of shapes that can cover a plane by continuously repeating a simple pattern, with what is called "translational symmetry". The most obvious examples are endlessly repeating squares or hexagons, which you may see in the set of tiles on a bathroom floor. An aperiodic tiling also covers the plane with a small set of shapes – but there is no individual section that is repeated over and over in a regular pattern. While the basic shapes appear infinitely often, there is no grouping of them that is repeated forever to fill the plane. Roger Penrose's "kites and darts" tiling, where two types of irregular quadrilaterals can together fill a plane in a non-repeating pattern, is one famous example. One other surprising aspect is that an aperiodic tiling may show rotational symmetry in local regions, where a set of tiles can be rotated by some angle and appear identical to the starting point, as in the 5-way symmetry of the original Penrose tiling.

Another unusual property of aperiodic tilings is that many of them can be obtained by projecting periodic lattices in 4 or more dimensions; in other words, the aperiodic nature is due to our low-D view of a 2-D "shadow" of a regular higher-dimensional structure. A nice online article by Stefan Weber illustrates graphically how this can be done, using a regular 2-D lattice to derive a 1-D aperiodic pattern. The key seems to be to place an n-dimensional hyperplane at an irrational slope in an $n+1$-dimensional structure, with the irrationality preventing its projection from having a repeating pattern. So in a sense, the fact that we see many aperiodic tilings as aperiodic can be considered simply artifacts of our limited view of higher-dimensional spaces.

How does this lead us to quasicrystals? Ordinary crystals are three-dimensional analogs of periodic tilings, where simple solid shapes are repeated endlessly to form the crystal. Well, given that I'm not currently embedded in crystal while typing this, "endlessly" isn't quite accurate, but I think you get the idea. Thus, it wasn't such a stretch for mathematicians to ask whether there could be 3-D aperiodic tilings that form another type of crystal. Physical scientists didn't think this was possible, but surprisingly, in 1984, a group of scientists reported an aluminum-manganese crystal whose diffraction pattern had a fivefold symmetry. This seemed very odd, since this form of symmetry was not thought possible for a crystal – similar to trying to fill a plane with pentagons. They concluded that it must be a case of an aperiodic crystal. The scientific community was skeptical at first, but this was followed by numerous other, similar discoveries. Finally in 1991 the International Union of Crystallography amended its definition of crystals to allow aperiodic cases.

Now there are hundreds of types of quasicrystals known, including some stable ones. In 2009, a natural quasicrystal was even discovered in a rock from a Russian mountain range. And some of the stable ones seem to have the very nice property, possibly due in part to their aperiodic nature, that it is very difficult for other materials to stick to them – making them an ideal material for coating nonstick cookware. A company called Sitram developed and began marketing a set of quasicrystal-coated cookware, known as Cybernox, in the last decade.

So, next time you want to make fun of mathematicians for spending time with crazy ideas of aperiodic tilings and higher-dimensional spaces, stop and fry yourself an egg first.

As Math Goes By
From Math Mutation podcast 134

Recently I've been reading the book *Warped Passages* by Lisa Randall, in which she talks about string theory, higher dimensions, and similar mind-bending aspects of modern physics. In the introduction to one chapter, she mentions that the famous song "As Time Goes By", which you have probably heard in the movie *Casablanca*, has additional verses that refer to Einstein and the fourth dimension. I assumed she was just pulling my leg, but did some web searching just in case. And I was surprised to find that multiple, seemingly independent, sources do confirm that the song has these lyrics.

The song, originally written in 1931 for a Broadway musical before being included in *Casablanca*, has a famous chorus. I'll spare you the pain of hearing me attempt to sing it, and just state the lyrics: "You must remember this / A kiss is just a kiss, a sigh is just a sigh. / The fundamental things apply / As time goes by." Makes a nice song, but seems pretty sappy, and doesn't have much to do with science or math. But the chorus, which is what made it into the movie and is usually heard in later covers, actually does not appear until the middle of the original song. The song begins as follows: "This day and age we're living in / Gives cause for apprehension / With speed and new invention / And things like fourth dimension / Yet we get a trifle weary / With Mr. Einstein's theory / So we must get down to earth at times / Relax relieve the tension."

How did references to, at the time, cutting-edge ideas in physics and math, make it into a set of Broadway lyrics? A nice article by a Brown University professor named Philip Davis provides some insight. As Davis explains, "In the 1920s and later, the newspapers were full of Einstein and his theory. He was the consummate genius. The term "relativity" was in the air, and as it became a buzzword..." In fact, these ideas were so prevalent in the public consciousness that half-understood notions of relativity were incorrectly applied to daily activities: for example, when a drive in the car didn't take the expected time, people would blame Einstein's theory. Unless they were traveling at a significant fraction of the speed of light, relativity was not truly a detectable factor. The author of "As Time Goes By", Herman Hupfeld, was a graduate of Cornell, and thus likely to be at the upper levels of comprehension in this area, at least as far as pop culture figures go.

A more interesting question is whether the themes of the song, and of the movie *Casablanca*, truly mesh with the references to Einstein and the fourth dimension. I think a case can be made for this, though the producers of the movie obviously didn't think so. Davis describes the connection as follows: "The times change and we change with them. Despite all the revolutionary changes that occurred in his lifetime, songwriter Hupfeld asserts that something in human nature abides. Mathematics and physics seek invariants; Hupfeld sought the invariants of human experience."

Amusingly, covers of this song by artists including Binnie Hale and Rudy Vallee changed one word: "fourth dimension" became "third dimension". Were these artists trying to make a profound statement about the flatness of modern human existence? Or were they just air-headed celebrities who had never heard of the fourth dimension and assumed it was a typo? We'll never know for sure.

Between the Dimensions
From Math Mutation podcast 22

We all know that a one-dimensional figure is a line or curve, essentially something we can measure just by finding the linear distance from a starting point. And a two-dimensional figure is a surface, where we need two coordinates, like a latitude and longitude, to figure out where we are. So what would a 1.2-dimensional figure look like? The question seems silly at first – who could imagine such an absurdity – but actually, figures with such fractional dimensions have been defined. In fact, these fractional dimensions are used to describe fractals, such as the Koch Snowflake we discussed in Chapter 2.

To review, a fractal like the Koch snowflake is essentially an infinitely detailed curve. (This description actually covers only a subset of fractals, but it is sufficient for this discussion.) These are useful to model things like mountains and coastlines. To understand the concept of being infinitely detailed, let's take a closer look at how a real-life coastline would be measured. Suppose you look at a map of the world, and try to measure the length of the Oregon coast. You probably would not be able to see many features at a detail level below 100 miles or so, and might come up with an estimate of about 300 miles. Now suppose you are just looking at map of Oregon. You can see lots of minor indentations in the coast that you missed at the coarse-grained scale... so it may look closer to 350 miles. Now suppose you are feeling energetic, and decide to walk along the beach from Astoria to Brookings with a yardstick, measuring as you go. At this level, you will follow many tiny bays and inlets too small to show up on a state map, and when you add all your distances up, will find it is even longer, maybe hundreds more miles. Then, if you crawl your way back with an electron microscope being sure not to miss a single grain of sand, you will extend the length still further.

So, how long *is* the coast of Oregon? The question doesn't really have a good answer – it all depends on your granularity of measurement. In real life, there is some level where you have to stop, of course. But a fractal is an ideal mathematical curve that represents this situation, where you can keep looking at a finer and finer granularity down to infinity, increasing the length each time. In other words, in fractals, there can actually be an infinite distance between any two points. For fractals in this class, if you ever claim you traversed a finite distance, and then look more closely, you will be guaranteed to find some outcropping or inlet in the curve that you missed. And a single measurement coordinate cannot describe where you are. Thus, fractals can be considered to have fractional dimensions.

A fractal like the Koch snowflake is a curve, not a surface, so it can't be said to be two-dimensional. But somehow it is more than one-dimensional, since a single coordinate (indicating distance from a designated origin point) cannot specify a point on it. Mathematicians have defined precise methods for calculating a curve's dimension, essentially depending on the speed with which increased detail appears when you look at a finer granularity. In the case of the Koch snowflake, it is considered to be approximately 1.26-dimensional.

Now you may object that we are cheating here, using the word 'dimension' to describe some new measurement that doesn't really correspond to our intuition about dimensions. I have to say, I can't totally disagree with that objection. Yet on the other hand, I think it is clear that our ordinary notion of dimension does somehow break down in the case of fractals.

CHAPTER 5

■ ■ ■

Understanding the Universe

Like many of you reading this book, I have been a lifelong fan of science fiction movies, books, and television shows. This is not very unusual for a math and computer geek, since there has always been a strong correlation between interest in mathematics, space travel, and science fiction. To some extent, this is inherent in the topic: after all, our space program and our knowledge of astrophysics, astronomy, and cosmology are heavily dependent on some of our most advanced mathematical techniques. But I think it goes beyond this, with some fundamental similarity between the way we stretch our minds to think about what might lie beyond us in space, and what might lie in the many mathematical possibilities beyond our mundane systems of counting and measuring.

The Bogus Bang?
From Math Mutation podcast 68

We're all used to the Big Bang theory by now, the standard model of cosmology that says our universe started out infinitely hot and dense, and suddenly expanded outward. You've also probably heard of superstring theory, which we mentioned in the last chapter. Among other things, this predicts that the universe actually contains 11 dimensions, not just the three we can see, and in fact we are living on some kind of gigantic multidimensional membrane, or "brane", floating within this higher-dimensional space. These two theories don't seem to inherently contradict each other, to a layman at least, but in fact there are a lot of details of the Big Bang, especially the way it describes the first moments of our universe, that have made some physicists nervous. Recently I was intrigued to read in Rudy Rucker's blog about a new theory, created a few years ago by physicists Paul Steinhardt of Princeton and Neil Turok of Cambridge, that uses this new view of the multidimensional nature of our universe to replace the Big Bang model with something a bit different.

The basic idea is that if we are living on a membrane floating around a multidimensional space, why should it be the only one? Suppose there are other, similar *branes* floating around. What happens when a pair collide? Steinhardt and Turok investigated this question, and spent over a year trying to work on the equations that would describe this situation. They discovered that when two of these membranes get close, they would begin to ripple and distort. Then at some point, the peaks of a pair of ripples would collide, and create an effect very similar to the Big Bang. The force of the

E. Seligman, *Math Mutation Classics*, DOI 10.1007/978-1-4842-1892-1_5

impact would cause a burst of energy and a rapid expansion of space, just as in the Big Bang theory, and due to the various peaks and valleys in the ripples, there would be hot and cold spots after the collision. These hot and cold spots correspond to what we actually observe in the universe. This new model is known as the *ekpyrotic* model, from the Greek word for conflagration.

What's even more interesting, though, is the long-term future in this model. Currently, with the expansion of the universe accelerating, the Big Bang theory predicts that our universe will slowly become essentially empty as all particles of matter accelerate away from each other. But in the new theory, the nearby neighbor brane that we originally collided with will still be out there, floating in multidimensional space – and eventually, there will be another collision between the two now-empty branes, that will form another Big Bang-like explosion and start the cycle of matter all over again. So, instead of just being on a long path to emptiness, this says we live in a cyclic universe, which will eventually renew itself and start again.

As with any new theory, though, there is quite a bit of controversy. Like the Big Bang theory, this theory still suffers from a singularity; that is, a point where its equations break down, at the very beginning of the universe. And even Steinhardt has a disclaimer on his web page that this idea is based on unproven ideas in string theory. There is work underway to try to find evidence one way or the other, in the form of gravity waves that would exist in the Big Bang but not the ekpyrotic theory, but results will be a long way off.

Still, it's fascinating to think that, as professor David Spergel of Princeton has said, maybe "everything that astronomers have ever observed is just a speck within the higher dimensions, and all of history since the Big Bang is but an instant in the infinity of time."

The Shape of the Universe
From Math Mutation podcast 86

If you're a *Math Mutation* listener and a fan of the TV Show *The Simpsons*, I'm sure you remember the episode where Lisa joined Mensa, and physicist Steven Hawking visited Springfield to meet her. At one point in the episode, Hawking has a beer with Homer, and tells him, "Your theory of a donut-shaped universe is intriguing... I may have to steal it." Like me, you probably paused to wonder if this joke was based on some real-life theory of the shape of the universe. The answer, as is so often the case on that show, is yes.

To start with, let's try to get our minds around the concept of the "shape of the universe". As we have discussed before, modern physics sees the universe as a multidimensional object, so we're really talking about some higher-dimensional analog of shapes, rather than the familiar 3-D shapes that we know. But we can get the basic idea here by looking again at the metaphor of two-dimensional creatures, or flatlanders, living on the surface of some 3-D universe. The simplest example is to think about the surface of a sphere: a two-dimensional creature traveling around a small part of it would imagine themselves to be on a large, flat plane, and would then be surprised to see that if they travelled east for enough time, they would come back from the west. So if our universe is on the 3-D surface of a 4-D hypersphere, we too might one day return to where we started, having gone "all the way around" this higher-dimensional surface.

While the sphere is the simplest metaphor, there are many other possibilities for the shape of the universe. The donut, or in math-speak the torus, is another possibility discussed by physicists. How would the universe appear different to a flatlander on a giant torus? Like on a sphere, if you travel long enough in one direction, you'll return to your starting point. In, say, the east-west direction, if you travel around the full circle of the donut, you would have to travel the entire distance of the universe, just like in the sphere. But in the other direction, say north-south, you would be traveling in and through the donut hole, having only to travel a much smaller distance before seeing your starting point.

But there's no reason why we should stop at the torus, and in fact physicists have many more bizarre shapes for the universe in mind. In the previous section we mentioned an alternative to the big-bang theory, where we are said to be on higher-dimensional wobbly membranes. Another idea has been proposed recently is a kind of dodecahedron connected to itself in higher dimensions. Think of ourselves inside a giant soccer ball, and if we exit it on one face, we find ourselves re-entering on the opposite side, but rotated 36 degrees. And an even stranger possibility that physicists have taken seriously is the idea of a giant horn shape known as a "Picard horn", with a narrow end getting narrower and narrower towards infinity, but the wide horn opening having a definite end; if a traveler reaches the end of the horn, he circles around and starts heading inside the interior. In this case, the universe is non-uniform: while it appears very large near the mouth of the horn, where we probably are, at the narrow end it gets arbitrarily small. So you could travel to a part of the universe where you are nearly as wide as the universe itself: you could reach out ahead and poke yourself in the back, your arm having wrapped all the way around the universe. In this case, you would appear to be in a gigantic hall of mirrors, looking at your own back, at a 2-universe-distance view of your own back based on reflected light that has wrapped around twice, etcetera on to infinity.

Physicists at the NASA WMAP project have been studying the universe's background radiation, hoping to get clues about the actual shape that we are living in. Recent results posted online seem to put a damper on the more bizarre theories, suggesting that NASA has concluded the universe as a whole is flat rather than curved, with only a 2% margin of error. (This is analyzing the global structure; there is still local curvature, as described by Einstein). But weird theories and radical shifts in conventional wisdom seem to be the one constant of the physics universe, so who knows what they will say a few years from now, or by the time this book goes to press.

Your Size in Space and Time
From Math Mutation podcast 148

Have you ever wondered about exactly how big you are in the scheme of things, when compared to the total size or duration of the universe? How does our extent in space compare to our extent in time? Since both length and time are dimensions we can calculate, it shouldn't be too hard to compute.

The observable universe is a sphere with a diameter of about 92 billion, or 9.2×10^{10}, light-years. A light year is about 10^{16} meters. Assuming you are about 2 meters tall, your height in universes is about $2 / (9.2 \times 10^{26})$, or 2.17×10^{-27} universes.

Now how about your size in time? Time is just another dimension for the purpose of this calculation, so we should treat it the same way. The universe is about 14 billion years old, though estimates do vary a bit. Let's take an optimistic view of medical science, and assume you will live about 100 years. Your size in time is then $100 / 1.4 \times 10^{10}$, or about 7.14×10^{-9} universes.

What I find surprising here is that your size in time is so much greater than your size in space, by a factor of about 3×10^{18}, or 3000000000000000000. That means in terms of your impact on the universe, the time you spend here is much greater than your physical extent. If your size in space were as large as your size in time, you would be a major astronomical object. And if your size in time were comparable to your size in space, you would appear and disappear in an instant. Strange to see how much greater our extent in time is than our extent in space, isn't it? Does it mean something about our significance for posterity as opposed to our significance right now?

I'm sure you have spotted a couple of problems with this calculation. For one, we don't know the real size of the universe: beyond the observable universe there may be an infinite extent of other stuff, too far away for its light to ever reach us. As for the size in time, we really should count the future as well as the past, and we don't even know for sure if the total duration will be finite or infinite – so we may actually be as small in time as we are in space, just happening to be located near the beginning.

You might also raise some more philosophical objections here: is our size in time just confined to our lifetime, or does it live beyond as we are remembered and continue to influence future generations? Should Alexander the Great or Julius Caesar get 2000 years of credit since we still remember their names today? Do I still get to count the years after I die, until some sysadmin at iTunes finally gets around to deleting this podcast, and anti-math barbarians burn all copies of this book?

I originally saw this calculation in Robert Grudin's book *Time and the Art of Living*. Grudin makes a lot of philosophical observations like this about our relationship with time and space. They may not always ring true, but they do make you think a bit.

Observing the Universe
From Math Mutation podcast 61

How big is the universe? There is no way for us to answer that question with certainty based on our current techniques – we need to start by talking about the 'observable universe', all the things we can actually detect. If something is so far away that light traveling since the creation of the universe would not have had time to reach us, then we are probably out of luck. So, let's start with that: how big is the universe we can actually detect?

You might think the answer is pretty simple at first. According to current theories, the universe is about 13.7 billion years old, so just draw a sphere of radius 13.7 billion light-years around the Earth, and that describes what we can see. But surprisingly, that answer is wrong. The problem is that we are in a universe that has been expanding in more than the three dimensions we are used to: ever since Einstein, we have realized that any calculations at the cosmic scale must utilize a four-dimensional structure of spacetime. Returning to a Flatland-like analogy, think of our space as the surface of a giant balloon that is being slowly blown up. Things that were once close enough to emit light that we can observe today may get farther and farther away as the balloon expands – so the light that was launched long ago can arrive at our telescopes, even though the actual object is now much farther away than that 13.7 billion light-years. As a result, our observable universe contains objects that are now as many as 46.5 billion light-years away.

An even more bizarre property of the observable universe is that we might actually be observing a *larger* universe than the one that actually exists, due to light that completely travels around the universe and reaches us a second time. In other words, the universe might be like the hall of mirrors at a carnival, with the same objects being visible to us many times. Using the balloon-surface analogy again, look at a line directly from a distant galaxy to us on the surface of the balloon, and also at another line that loops all the way around the balloon and then to us. Both might be legitimate paths for the light to reach us, and we would see them as separate images. A question you might ask now is: why are we so gullible? Shouldn't we notice immediately if some object in the sky is exactly the same as some other one? The problem here is that since the light must likely traverse many billions of light-years to go all the way around the universe, the duplicate objects will be viewed at vastly different eras in their history. It's not easy to know exactly what a particular galaxy will look like in several billion years, or what it looked like that amount of time in the past. Some object we have labelled as a "distant galaxy" might actually be our own Milky Way, viewed many billions of years ago.

Of course, we also need to keep in mind the very likely possibility that the universe is much larger than what we can detect. This discussion has by necessity been limited to the observable universe, since that's all we can realistically comment upon. But it is very likely that we really are observing only a subset, and we may never know how big the universe really is.

Alien Algebra
From Math Mutation podcast 129

A listener recently emailed me an interesting question: He had heard that we could theoretically communicate with aliens using mathematics. But clearly there are many symbols and conventions of math that are culturally determined, not universal: how would an alien know what a plus sign is, or what an equal sign is, or identify that squiggly thing we use to symbolize an integral in calculus? But the answer is a pretty simple one: While it is true that there are many such aspects to modern math, there are in reality many universals that we could take advantage of for communication, even without any shared culture.

One basic universal that we might be able to take advantage of is the concept of prime numbers. This method has been suggested by numerous scientists and mathematicians, and is probably most famous from its appearance in Carl Sagan's novel *Contact*. As you may recall, a prime number is a number greater than one which is divisible only by itself and 1. Due to the fact that any number is a product of a unique set of prime factors, regardless of base, units, or measuring system, the concept of primes would surely have been discovered by any species advanced enough in mathematics for interstellar communication. It doesn't matter if there are 5-headed hydras on Alpha Centauri who count using their 14 tentacles in base 62: the concept of prime numbers would still exist, and the same numbers would be prime. In addition, the known algorithms to determine prime numbers are complex enough that it is very unlikely that some natural process would arise by chance and generate them. So, if we were to send a set of pulses into space that first repeated twice, then three times, then five times, and so on, expressing the first 100 prime numbers, anyone detecting them would be very likely to assume they were generated by some kind of intelligence.

While sending pulses of prime numbers might be a nice way to establish the existence of intelligence, it's hard to see how this could lead to any useful communication. How could we move from simply being aware of distant aliens' existence to actual messages with content? The problem seems insurmountable at first. But surprisingly, in the 1960s, astronomer Frank Drake came up with a clever scheme that could actually be used to establish real communication with distant, unidentified aliens.

Here's how it works. Take two reasonably large prime numbers, which we'll call p and q, and construct a rectangular video display that measures p pixels by q pixels. Each dot of the display can be either on or off. If we then take all the dots of the display, representing dots by a short pulse if off and a long pulse if on, we can convert them into a long linear series of pulses of length p × q. We can then repeatedly send this message into space.

How would the aliens know what to do with this message? Well, hopefully they would recognize that this repeating series of pulses has a length which is the product of two large primes, especially if we have already established our existence by broadcasting prime number series. Assuming that the primes we choose are relatively large, the chance of a natural process randomly producing repeating pulses of that size would probably be very small. Thus the alien scientists would suspect an intelligent origin. They should then realize that this message, as the product of precisely two primes, could be displayed as a two-dimensional array, either p across and q down, or q across and p down. One of these possibilities would replicate our original p by q display!

Using this method, we could transmit black-and-white pictures of arbitrary resolution. We could send every photo on Facebook to the aliens, though that might be a bad idea if we want them to return our calls. But we could also send them examples of our biology, our arts, and or technology. We could even talk about particular elements, using sets of *n* dots to represent the element with atomic number *n*, and if they are clever about interpreting the pictures, this might even inform the aliens of the chemical content of the objects in the pictures we send.

So, using some basic math, we could establish a surprising level of communication with alien species. In fact, we are already trying to do it, with SETI (the Search for Extraterrestrial Intelligence) broadcasting an image using the Drake method that contains pictures including people, the double helix of DNA, representations of the atoms making up DNA, and schematics of the radio telescope broadcasting the message. Unfortunately, we have to be a little patient, as odds are that even if someone receives it, the reply will not arrive for thousands of years – if we're still around to recognize it as a reply.

Time Reversed Worlds
From Math Mutation podcast 195

Recently I read Martin Gardner's classic book *The New Ambidextrous Universe*, on various forms of symmetry found in modern physics. One of the most amusing ideas discussed in this book is the idea of time-reversed worlds. Is the direction of time truly fixed, as it seems to be from our point of view? Or could there be parts of our universe where time runs in reverse from how we observe it, so our future is their past, and vice versa? Naturally, this idea has been explored by many science fiction writers over the past century. I think my favorite use of the idea was by Kurt Vonnegut in his novel *Slaughterhouse-Five*. Let me quote a bit from Vonnegut's description of how a time-reversed observer would describe the bombing of Dresden during World War II:

The formation flew backwards over a German city that was in flames. The bombers opened their bomb bay doors, exerted a miraculous magnetism which shrunk the fires, gathered them into cylindrical steel containers, and lifted the containers into the bellies of the planes. The containers were stored neatly in racks. ... When the bombers got back to their base, the steel cylinders were taken from the racks and shipped back to the United States of America, where factories were operating night and day, dismantling the cylinders, separating the dangerous contents into minerals. Touchingly, it was mainly women who did this work. The minerals were then shipped to specialists in remote areas. It was their business to put them into the ground, to hide them cleverly, so they would never hurt anybody ever again.

More seriously, we might ask the question, how could a time-reversed world be possible? One idea comes from the concept of antimatter. As you may recall from physics, most fundamental particles have a corresponding antiparticle with opposite charge. For example, an electron has a negative charge, and its antiparticle the positron has a positive charge. When a particle and an antiparticle collide, they annihilate each other in a burst of energy, creating a new photon, or light particle. As we discussed in the last chapter, Feynman had a curious insight. Perhaps antiparticles could simply be particles traveling backwards in time. For example, a typical interaction in a Feynman diagram might show an electron and positron colliding, generating a short-lived photon, and the photon then creating an electron-positron pair. But you could also interpret this as the first electron suddenly reversing direction, emitting a photon as it turns backwards in time to become a positron. Then a short time later, a positron traveling backwards in time collides with the photon, and reverses direction to become a forward-moving electron. The resulting picture is exactly the same.

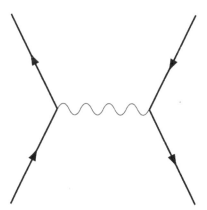

Figure 5-1. *Feynman diagram of electron-positron collision*

If antimatter is just matter traveling backwards in time, could there be entire solar systems and worlds somewhere out there in the universe that are made of antiparticles, and thus experiencing time backwards from our point of view? It's initially challenging to tell directly if a distant solar system is matter or antimatter, since the photon is its own antiparticle, and thus an antimatter solar system would look just like a matter one through a telescope. However, modern physicists have realized that interstellar space does contain some atoms, approximately one per cubic meter. This doesn't sound like a lot, but is enough that any matter/antimatter region boundary should be detectable due to the particles annihilating each other and generating gamma ray bursts. There are also some heavier particles that occasionally arrive in cosmic rays, and these suggest where we can also look for evidence of antimatter regions. The physicists looking for these phenomena are now pretty sure that there are no antimatter regions in the observable universe. However, there could still be such regions in areas too far away for us to observe.

As you would expect, there are other objections to the possibility of time reversal in our universe, such as the need to follow laws of entropy, which seem to point time monotonically in one direction. And if some time-reversed antimatter aliens stopped by to visit, it would be a rather unpleasant encounter, with them getting annihilated particle by particle as soon as they arrived. We might try to communicate with them by radio, but it would be a rather confusing conversation, as the only questions they could answer would be ones that we hadn't transmitted yet. But perhaps by programming long-lasting computers to send messages sometime in the future, we could at least have a rudimentary discussion and become aware of the basics of each other's existence.

A Pear-Shaped Planet
From Math Mutation podcast 120

If you grew up in the U.S. before the mid-1970s or so, you probably learned in elementary school that Christopher Columbus believed the Earth to be round, while his foolish contemporaries thought it was flat. Thus Columbus was a visionary who first understood the true shape of the Earth. But probably a few years later, you heard the correct story, that our planet had been known to be a sphere since antiquity, and that the dispute was merely about the size. The flat Earth story had been part of a 19th-century romanticized Columbus biography by Washington Irving, and entered the popular culture from there.

Actually, not only did the ancients already believe the Earth to be spherical, but in the 3rd century BC, Eratosthenes came up with a remarkably accurate estimate of the Earth's size. He did this by observing that when the sun is directly overhead at Syene, as seen by lighting the bottom of a straight well, it was offset at an angle equal to 1/50th of a circle in the city of Alexandria, which was directly north. Since this offset angle should be theoretically equal to the portion of the Earth's arc covered by the distance between the two cities, he then multiplied the distance between Syene and Alexandria by 50, and came up with an estimate of about 25000 miles.

Interestingly, *Wikipedia* points out several aspects of this story that are suspicious: the cities aren't exactly on a north-south meridian, the angle measurement is a bit off, and we don't really know the exact size of Greek 'stadia' units in terms of today's units. So the accuracy of Eratosthenes' estimate of the Earth's size may in fact be an urban legend. But it's generally agreed that the ancient Greeks did know the Earth to be a sphere, and got a measurement in the right ballpark.

Columbus was aware of this information, but decided to put more trust in some other estimates of our planet's size. A scholar named Marinus of Tyre, around the end of the 1st century AD, had come up with estimates that the Eurasian land mass took up about 225 degrees of the Earth's surface, making the planet much smaller than other ancient estimates. We shouldn't be too hard on Marinus – he did also make the contribution of first providing latitude and longitude to identify locations on a map, an invaluable tool for future mapmakers.

But as you have probably heard, when Columbus encountered America, he took this as confirmation that Marinus' smaller estimate of the Earth's size was accurate, and assumed he had reached East Asia. Most stories of Columbus end there, with him deluded until the end of his days thinking that he has proven the theory of the smaller Earth. But there is one more wrinkle that you may not have heard. In his later years, Columbus saw a lot of data from various other voyages, including his own trip to South America on his third voyage. This additional data, plus the quantity of fresh water pouring from rivers into the Atlantic, convinced him that he had encountered another vast continent, contradicting his earlier confirmation of the Earth's small size.

How did he reconcile this with his beliefs? He decided that the planet must not be a sphere at all, but had to be pear-shaped, with a big lump stretching out somewhere near China to make room for a big new land mass. Back in Europe, he was widely mocked for this bizarre idea. Ironically, we know today that he was right that the Earth is slightly pear-shaped, though bulging in a different way than he expected: the southern hemisphere is a tiny bit 'fatter' than the northern. But it's a relatively tiny difference compared to what Columbus estimated, so he isn't fully vindicated by this modern knowledge.

Where Am I?
From Math Mutation podcast 108

In these modern days of satellite photos and global positioning systems, we take it for granted that whether we are at home, in a boat, or on an airplane, we can pretty much always know where we are on the planet. Two numbers always describe our location: the latitude, saying how far north or south we are, and the longitude, specifying our location east and west. But have you ever wondered when it was that people gained the ability to calculate these numbers while traveling?

In particular, on a boat in pre-modern times, when you could see nothing but water in every direction, knowing your latitude and longitude was a life-or-death matter. Surprisingly, although latitude could be calculated since ancient times, the problem of longitude wasn't solved until the 1700s. This asymmetry might seem a bit odd at first, but if you think about it for a minute, I think you'll see the problem: while people at two different latitudes will observe clearly different views of the stars at night, people at different longitudes will see the same views, just at different times, due to the rotation of the earth.

So how was the problem of longitude solved? As long ago as the second century BC, the Greek astronomer Hipparchus recognized that if you could accurately calculate the current time at a known location and compare it to the time of day at your present location, that would give you a precise measure. But it's no easy matter to construct a mechanical clock that still works in the chaotic, constantly-moving conditions on the ocean, and was suspected by many to be an impossible dream. For a long time, sailors were limited to "dead reckoning", or calculating based on estimated speeds of a ship how much time had passed. This often led to accumulated errors, and extra days or weeks at sea meant increasing risk of shipwrecks, starvation, or death by scurvy. Recognizing the often fatal consequences of miscalculations, various European governments began offering lucrative prizes in the 1600s and 1700s for an accurate way to measure longitude.

The British actually created a government body, the Board of Longitude, to examine the various solutions and administer the large reward. Needless to say, like in modern patent offices, this caused all sorts of crackpots to come out of the woodwork, convinced that they were geniuses who had saved the world. Someone named Owen Straton attempted to claim the prize by presenting a sundial. The somewhat impractical Halandby solution involved making sure that you are always traveling precisely at a 45 degree angle northwest, southwest, northeast, or southeast, regardless of ocean current, and then simply reusing your latitude calculation. Lucasian professor William Whiston proposed another impractical scheme, permanently anchoring ships at regular intervals across the Atlantic and having them launch fireworks at known times. Probably the most ridiculous idea was one involving the semi-magical "powder of sympathy", where a bandage that had previously been on a wounded dog would be poked at certain times, and the dog would then feel the wound again no matter how far away he was. Umberto Eco mocked this idea in his classic novel *The Island of the Day Before*.

But among all the clutter were some practical methods. Galileo had noted in 1612 that due to the regular orbit of Jupiter's four moons, Jupiter itself was a kind of universal clock: by observing the relative positions of the moons, you could always know the exact time. This worked well on land, but observing the moons of Jupiter from the deck of a 1600s-era ship was nearly impossible, even with the clever apparatus invented by Galileo, which attempted to use layered shells separated by oil to enable an observer to stand still on a moving vessel.

Related to this method is the method of lunar distances, using the exact position of the Earth's moon to identify the current time. This was proposed as early as 1514 by Johannes Werner, but did not really become practical until the release of Nevil Maskylene's *Nautical Almanac* in 1767, which contained detailed tables of lunar positions. It still suffered from the labor-intensive need to make a local observation of the moon's position with respect to the Sun or a known star, but to some degree was a practical method.

The problem was truly solved by British carpenter John Harrison, the first to come up with a practical chronometer that could give the exact time even at sea. Harrison was a self-educated clockmaker, said to have first learned of clocks at the age of 6 while amusing himself with a watch while sick in bed with smallpox. He truly mastered his craft; according to online sources, one of his original clocks from the 1720s still works today. He created numerous innovations in clock design, such as systems of counter-weights to compensate for nautical motion, and the gridiron pendulum and bimetallic strip to compensate for temperature changes. He first demonstrated a successful marine chronometer in 1736. He was given various interim grants to continue work on it, but to his misfortune Nevil Maskylene, famous contributor to the lunar distance method, was appointed Astronomer Royal before Harrison could formally be granted the Longitude Prize. Maskylene sabotaged the official test of Harrison's 'H4' chronometer, ironically the same design successfully used in practice by the famous explorer James Cook, by distorting his presentation of the results. Eventually the elderly Harrison's accomplishments were recognized by an Act of Parliament in 1773, though he was never granted the official longitude prize.

Because Harrison's chronometers were initially expensive, the labor-intensive and less-accurate, but still useful, lunar distance method remained popular through the mid-1800s, only gradually being replaced as the price of chronometers descended to convenient levels. And today we're all familiar with the various modern technologies that eventually made the chronometers themselves unnecessary, despite their status as marvels of pre-modern mechanical technology.

Putting the Multiverse to Work
From Math Mutation podcast 31

Quantum Computing is a new form of computation that can turn many of our ordinary assumptions about solvable problems on their head. For certain types of computations, quantum computers enable the performance of an exponential amount of computation in polynomial time. When we say a computation requires "exponential" work, it means the amount of work required to solve a problem of size n is roughly proportional to 2^n. Since there are estimated to be only about 2^{260} atoms in the observable universe, this means that such problems quickly grow unsolvable for most practical purposes. So how can a new form of computer address this issue?

To get the basic idea, let's review the concept of bits, the 1 s or 0 s that represent the information in a standard computer. If you have a 2-bit memory element, it can have one of four values: 00, 01, 10, or 11. But in a quantum computer, the basic elements are something called *qubits*. Qubits can be in a *superposition* of multiple states at once: so rather than one of those four definite states, a 2-qubit memory element is in a state of A times 00 + B times 01 + C times 10 + D times 11, where A, B, C, and D are complex probabilities. A set of n qubits thus represents a set of 2^n coefficients, an exponential amount of information. To perform computations with such elements, the laws of quantum physics are used to cause the full sets of coefficients of different memory elements to interact, eventually resulting in a superposition that represents a correct answer with a probability close to 1.

So, does this mean all our problems of computation are now solved? Well, not quite. The set of physical interactions that are available in quantum computers is very limited, so arbitrary exponential problems cannot be solved: only certain ones that fit within the possible interactions. One problem that has been proven to be solvable on such computers, though, is the factoring of large numbers – and many modern encryption algorithms, including the popular "RSA" algorithm, rely on the fact that such factoring is hard. As a result, if a practical quantum computer were suddenly on the market, many of our computer security systems would be immediately compromised! But you don't need to cancel all your electronic accounts quite yet, since nobody has built a proven quantum computer with more than a handful of qubits.

What I find most interesting about quantum computers is that physicist David Deutsch has argued that they support the "many-worlds interpretation" of quantum physics. The many-worlds interpretation holds that the superposition of many possible states of a quantum particle really represents the fact that there are many parallel universes, an exponential number, and in each one the particle holds a different state. The nearby parallel universes are just like our own, except for tiny quantum variations – there are an exponential number of other versions of you reading this book right now.

Some of them are even enjoying it. This concept of multiple parallel universes does seem a bit hard to swallow, especially given the large number involved. Martin Gardner famously ridiculed this idea when he asked, "Are universes thicker than blackberries?"

However, looking at the issue from the other direction, what would it mean if a successful large-scale quantum computer were produced? Deutsch points out that it could be taken as a strong argument that the existence of multiple universes has been proven. In this interpretation, the equations of quantum physics that describe the state of a particle represent the interaction between our universe and the many nearby parallel universes in which that particle exists. For a quantum computer to solve a practical problem that would require an exponential amount of classical work, the work has to be done somewhere – how can a simple particle be doing an exponential amount of computation on its own? If a quantum computer really succeeds, that could provide evidence that an exponential number of universes has been harnessed in order to perform our computation.

I'm sure I'm not doing Deutsch's argument justice here, but would encourage you to read more in his excellent book *The Fabric of Reality*.

CHAPTER 6

■ ■ ■

The Mathematical Mind

Mathematics is one of the few purely mental activities that we humans can undertake; while we tend to use it as a tool to analyze the world around us, it really takes place entirely in our minds. This observation is especially important when trying to understand the world of *autistic savants*, people with impaired mental function in some areas who seem to be mathematical geniuses in others. You will see several articles here that relate to these savant thought processes, which have always intrigued me.

Beyond this, there are questions of how to understand our minds in general: are we just a set of mathematical functions floating around our brains? How does a blob of cells like our brain manage to understand mathematics anyway? There are also many ideas about how to use mathematical insights to improve all our thought processes, through changing our use of language or through other means. In this chapter we will examine these various insights into our mental workings from several directions.

What Color is this Podcast?
From Math Mutation podcast 6

I was recently reading online about 'synesthesia', a phenomenon where the senses get mixed up. For example, someone may claim that certain piano notes trigger various smells, or that seeing certain shapes result in an odd taste in their mouth. One possible form of this condition is when people claim that they "see" each number as a particular color. For example, to a synesthete, all 2s might appear to be green, while all 3s appear to be blue.

Like most of you, I was kind of skeptical when first hearing this. I figured that this was just a case of pretentious artsy types trying to sound weird and mysterious, or trying to be poetic. This opinion is reinforced by the fact that one of the most famous synesthetes was the author Vladimir Nabokov, a favorite among self-proclaimed intellectuals and literary academics. But I recently read in *Scientific American* online about a clever experiment that seems to disprove this theory, and show that some people really do see each number as a different color.

Here's how it works. The scientist prints out a sheet of paper that is mostly 5s, but with a few 2s mixed in. The 2s are set up so they form some recognizable pattern, such as a triangle. Normal people will find it very hard to distinguish the pattern at a glance, and will have to stare a long time picking out individual numbers. But to a synesthete,

E. Seligman, *Math Mutation Classics*, DOI 10.1007/978-1-4842-1892-1_6

because the 2s and 5s are different colors, the pattern is obvious, and they should spot it immediately. If they can pass this test, a synesthete can plausibly claim that they are really seeing the numbers as different colors, and not just consciously choosing to describe them with figurative language. In numerous trials, a majority of the self-proclaimed synesthetes actually passed this test.

So, the conclusion: synesthetes really do experience the world differently. Those artsy types claiming to be synesthetes are not just pretending to be weird and mysterious – in at least some cases, they *are* weird and mysterious. And to some of them, every number really does have its own color.

Computers on the Brain
From Math Mutation podcast 48

Now that we have established that synesthesia is real, and not just an attempt at metaphor or poetry, another question arises: are there cases where this could be powerful enough to provide a significant advantage in a science or engineering career? If a small set of people have different parts of their brains cross-wired somehow, where different senses are somehow associated with numbers, letters, or concepts, it stands to reason that there may be cases where this can function as a powerful practical tool.

Along these lines, I was surprised to read in a recent issue of *Chip Design* magazine about another form of synesthesia: certain engineers see different parts of computer chip schematics in various colors. I should probably give a little background first, for those of you who don't help design computer chips for a living. When designing a microprocessor, engineers use different types of basic symbols, or "gates", to represent sets of transistors that perform common tasks. One example is the AND gate, which has two wires coming in, and one wire coming out. The wire coming out will have a value of 1 if both its inputs were 1, and otherwise it will have a value of 0. Similarly, an OR gate drives a value of 1 if either of its inputs was 1. A NOT gate, or inverter, is even simpler, always outputting a 1 if its single input was 0, and vice versa. When viewing a design schematic, an engineer will typically be staring at a sea of symbols representing AND, OR, NOT, and more complex gates.

According to this article, a synesthetic engineer who was interviewed can see AND gates as yellow, and OR gates as green. At first one might suspect that rather than reacting to the logical function of the gates, he is simply perceiving the shapes used to represent the gates, since a well-known form of synesthesia associates shapes with colors. But the engineer insisted this was not the case – for example, the symbol usually used for an inverter is a triangle with a small circle at the end. But triangles can also appear as parts of more complex symbols, and he doesn't see the same colors in those cases, or when the same shapes just happen to be on a piece of paper. So somehow, the part of his brain that understands the logic of chip schematics seems to be cross-wired with the part that perceives colors! This can be quite useful for an engineer working with schematics.

If you think about it, this also might add a new dimension to our understanding of computation and our brains. I had always thought that numbers and shapes are directly inherited from the natural world, so it's not too surprising that our brain can directly exhibit odd behaviors like synesthesia related to these. Something about these concepts seems to be fundamentally a part of nature; even if humans had never evolved, numbers and shapes would still exist somehow. But the logic gates used to design computers seem like a complex human invention - sure, they model basic forms of mathematical logic, but in a way not easily found in nature. So it seems to me that the fact that these forms are also subject to synesthesia says that, somehow, the concepts of computation really are as basic to our brains as numbers.

The Rain Man's Secret
From Math Mutation podcast 10

You have probably heard about autistic savants, sometimes called 'idiot savants' by the politically incorrect. These are people who can barely function in the real world, but somehow manage to do amazing levels of mathematical calculations in their head. For a long time, it was a complete mystery how these people managed to perform their mental feats. But recently a highly functioning autistic savant named Daniel Tammet has written a fascinating memoir, where he describes his methods. Tammet has been called the "Rosetta Stone" of savants, since he is one of the few who has these talents, but also is able to communicate well enough to describe how he does it. The key lies in the phenomenon of synesthesia.

Basically, Tammet's brain is hard-wired to perceive numbers in a different way: when he thinks about or reads a number, he experiences shapes, sounds, and smells unique to that number. It's not that he makes an effort to visualize these aspects, or is remembering earlier experiences – these senses are real to him. As he describes his perceptions of numbers, "Each one is unique and has its own personality. The number 11 is friendly and 5 is loud, whereas 4 is both shy and quiet – it's my favorite number, perhaps because it reminds me of myself." He once told David Letterman that he reminded him of the number 117. Later Tammet goes on to write that he has visual and sometimes emotional responses to every number up to 10,000.

Now how does this help him with arithmetic calculations? Well, as he describes it, he can multiply by visualizing the shapes of two numbers, which then change and a third shape appears. He can read the answer off the new shape. Division somehow produces rotating spirals in his head, which he can read accurately to almost 100 decimal places. Prime numbers have a smooth, round, "pebble-like" quality, which enables him to recognize all primes up to 9,973.

What's even more interesting is how he uses numbers as a substitute for emotions, which most autistics find hard to understand or relate to. When a friend is sad or depressed, he thinks of the number 6, which he describes as "tiny black dots." To understand the concept of being intimidated by something, he imagines himself standing next to the number 9, which is very tall. When he is unhappy or anxious, he counts in his head, and the patterns formed by the numbers help to reassure him.

A reasonable person might ask if Tammet is just making all this stuff up. But he is a pretty credible source: he once recited π to 22,514 digits under controlled conditions, setting a European record. Unfortunately, assuming he is telling the truth, it looks like his abilities are beyond the reach of the rest of us. But they certainly do make for a fascinating read: if you enjoy thinking about this topic, be sure to check out his bestselling memoir, Born *on a Blue Day*.

Look Him in the Eye
From Math Mutation podcast 119

Daniel Tammet's book, discussed in the previous section, actually appears to be just one representative of a small cottage industry of autism-spectrum memoirs. Recently I read another one: *Look Me in the Eye*, the autobiography of someone named John Elder Robison with the mild form of autism known as Asperger's syndrome. Despite, but also because of, his condition, Robison led quite a colorful life: bouncing from high school dropout, to KISS roadie, to professional engineer, and finally settling down as an auto mechanic and entrepreneur.

As I mentioned, Robison suffers from Asperger's syndrome, a mild autism-spectrum disorder. He always found it difficult to emotionally connect with other people, to figure out what to say in social situations, or to look others in the eye during conversation. He tried to deal with this in several ways. At the simplest level, he tried to come up with mathematical algorithms to tell him what to say in response to others' comments. This often worked but sometimes led to embarrassing results. For example, in one incident, discussions of someone having an affair triggered his "ask names of unidentified persons" response, and generated seemingly rude and nosy questions.

But he also found an unusual way to cope, which led him to be a little more socially popular than many others with his condition: practical jokes. When you think about it, it makes a kind of sense – he could not figure out how to do or say what was appropriate in most social situations, so he was better off intentionally doing things that were absurdly inappropriate. Some of his classics included convincing his parents' friends at a party that he was a corrupt garbage man collecting payments for the mafia, and arranging for several tons of gravel to be delivered from a quarry to a teacher's driveway.

Like many with his condition, he also had some strong mathematical talents that developed from an early age. After discovering musical equipment, he soon realized that he had a gift for understanding how sounds were produced by amplifiers: he could visualize the sound waves created by various electronic components, and thus could modify an amplifier to create desired sound effects. This led to connections on the local musical scene, and paying jobs working on sound equipment after he dropped out of high school. Eventually he was discovered by the rock band KISS. Yes, *the* KISS, the guys with the crazy makeup. He created numerous special-effects guitars for them, using his electronic gifts. But he eventually realized this wasn't a very stable lifestyle.

When he applied for a corporate job, his amazing insights into electronics impressed Milton-Bradley enough that they hired him as a design engineer, despite his lack of a formal education. He was quite successful as an engineer, but after being promoted to management realized that he wasn't really suited for the corporate world. He moved back to his hometown and started an auto-repair business. Again his gifted insight into electronic and mechanical issues paid off, as he became well known as the mechanic of last resort who could resolve issues in high-end foreign cars that stumped most standard service providers.

I think the most interesting thing about reading biographies like Robison's and Tammet's is the realization that there is really a kind of continuum between normal, or 'neurotypical' people in the current lingo, and people with Asperger's and more severe forms of autism. I think the characteristics often labelled as 'nerdiness' are somewhere along this line: while I am not suffering from Asperger's or anything like that, I could recognize some of my own past struggles when Robison described his challenges in dealing with social situations. And I think any of us in mathematical, scientific, or engineering fields have found ourselves working or studying with people who are somewhere near Robison's point on the autism spectrum. I highly recommend reading Robison's and Tammet's books to gain better understanding of our colleagues, and help realize that underneath all that odd behavior they are not really that different from the rest of us.

Savants Are People Too
From Math Mutation podcast 97

Recently I've been reading Daniel Tammet's second book, *Embracing the Wide Sky*, a fascinating account of recent ideas and research on how the mind functions, peppered with insights from Tammet based on his own unusual thought patterns. One important aspect of this new book is his strong belief that autistic minds like his are not fundamentally different from other people's – he uses the same facilities as everyone else, just with certain parts stronger and other parts weaker than the general population. In particular, he vehemently disagrees with the notion of autistic savants as "human computers".

For example, many of us got our ideas about people like Tammet from the movie "Rain Man", which exposed the idea of autistic savants to a wide audience. In one famous scene, Dustin Hoffman's autistic character sees someone drop a box of matches, and he instantly counts how many matches are on the ground. This was inspired by an account from Oliver Sacks' psychological memoir, *The Man Who Mistook His Wife for a Hat*, where he interviewed a pair of savant twins, who indeed stated the number of matches that fell out of a matchbox. But Tammet points out that if those matches were on the twins' table, they were very likely to have already known how many are in the box: even if the number was not labelled on the outside, they could count them at their leisure, not an unlikely activity for them. Also, if over 100 matches fall down, it's virtually guaranteed that the view of some would be obscured, making it highly unlikely that even a sophisticated computer would be able to accurately give the number by viewing the fallen matches. In addition, no controlled scientific study has ever found a savant to have this instant-counting ability, despite their having demonstrated many other abilities.

Tammet also points out that in his mind, the number 111, which was how many matches Sacks' twins counted, is a very beautiful number: the natural symmetry of the three ones, the fact that it's divisible by three, and the fact that it is a multiple of 37 are all very interesting. With his synesthesia, which you will recall is his ability to 'see' numbers as shapes and colors, 111 is "full of beautiful bright white light". So Tammet believes that the most likely explanation is that the twins counted the matches at some point before the interview, and had no problem remembering this especially interesting number that corresponded to the count.

In addition, Sacks talked about other incredible math feats by the twins: another notable example is their game of mentioning high prime numbers to each other. He claims to have confirmed the accuracy of their primes up to 10 digits by checking in a mathematics text he had that listed the primes. But a book that included all primes up to 10 digits would contain over 400 million numbers! When faced with this criticism, Sacks replied that he had since lost the book he was using, so his claim conveniently could not be checked. Overall, Sacks seemed especially predisposed to consider the autistic twins as something vaguely nonhuman, describing them as "a sort of grotesque Tweedledee and Tweedledum". Setting out from the beginning to depict them this way, it's not surprising that he viewed everything they did as further confirming their strangeness. Tammet is obviously offended at this characterization, as he should be.

With regard to his personal accomplishments, Tammet points out that his π-reciting feat was not something fully auto-computed by his subconscious mind, but the result of several intense weeks of constant studying and practice. His synesthetic abilities certainly helped, but they did not replace the requirement to carefully think about the mathematics he was attempting. So while being brilliant and having an excellent memory plus off-the-chart synesthetic ability, Daniel Tammet is still a man, and not a machine.

The Uninhibited Brain
From Math Mutation podcast 99

As we discussed in the previous section, in his second memoir *Embracing the Wide Sky*, Daniel Tammet makes the case that he is not a "human computer". This still leaves the question open, however, of how to explain his strange condition. He has a theory for that: savant syndrome results from a lack of inhibition in communication between parts of his brain. In other words, pieces of the brain that are kept separate from each other in normal people are somehow in constant communication for people like him, and this results in his exceptional mathematical abilities.

To begin to understand this, let's review the first part of his explanation: synesthesia. This means that when he sees or hears a number, he also sees associated shapes and colors in his mind. For example, he describes prime numbers as having a smooth, pebble-like quality, and David Letterman reminds him of the number 117. You can see why this would help give him an excellent memory for numbers and their properties: one common mnemonic technique is to try to visualize pictures associated with things you want to remember, and for him this comes naturally with numbers.

But the most novel part of Tammet's thesis is that this synesthesia is not the root cause of his abilities, but a symptom of the deeper root cause, the lack of inhibitions between parts of his brain. For example, think about what happens in a normal mind like yours when you hear the word "giraffe". A whole flood of associations come to you: the visual image of the animal, concepts like "tall", "mammal", and "long neck", etc. With numbers, things like this don't happen: when asked what comes to mind when you hear "23", for example, you may think of 22 or 24, but won't come up with many interesting semantic relations.

For Tammet, the ordinary associative abilities that people naturally have with language also extend to the mathematical parts of his mind. When he hears 23, all sorts of things immediately pop into his head with no effort: its square, 529; its largest 3-digit multiple, 989; and the fact that it's prime. Just like a normal person has an intricately networked linguistic vocabulary of thousands of words and concepts that they can associate without effort, Tammet has the same situation with numbers, his basic 'vocabulary' covering all numbers under 10000, and a nice selection of interesting ones beyond.

How does this let him perform new math tricks, like figuring out whether a 6-digit number is prime? The key is that he uses his connected insights to infer additional properties. He gives the analogy of finding words in a game of Scrabble: if you have the right tiles, you might guess that "agedness" is a word and play it, even though you have never seen that word written. Similarly, he might find a number that 'feels' prime, think about it a bit, and tentatively conclude that he has found a new prime number. Just like recognizing that his base word 'aged' plus the suffix 'ness' is probably a legal combination, he can split up parts of a number and figure out if they are combined in ways that have special meaning. For example, look at 84,187. He can instantly see that the first 3 digits make 841, or 29×29, and the last two make 87, or 3×29, and thus instantly recognize that 84,187 is not prime. To recognize a large prime number, he looks at its 'shape' for a few minutes in his mind, and sees if any factors occur to him: if not, it's probably prime.

An interesting consequence of this, and another refutation of the idea of savants as human computers, is the fact that it's not uncommon for savants to incorrectly label a large number as prime, if its factors are hard to find. For example, when asked to find a prime between 10,500 and 10,600, one savant came up quickly with 10,511, which 'looks' prime to a savant's mind, not having any obvious factors, or any immediate structure the reveals divisibility by some smaller number. But 10,511 is actually the product of 23 and 457.

So, if Tammet's theory is correct, a large part of the secret of autistic savants is that the parts of the brain handling their number sense and language sense kind of flow together, causing numbers to bring the floods of associations that others typically experience as parts of language processing. While you may be jealous of such abilities, watch out – this may also be a partial explanation of the autistic issues in their psychology as well, as the sensory overload is simply too difficult to process for a normal mind. And unfortunately, the majority of autistics with these savant abilities are not lucky enough to be as high-functioning as Daniel Tammet.

A Logical Language
From Math Mutation podcast 192

If you're a speaker of English, which is pretty likely since you're reading this book, you may have found yourself occasionally frustrated by its arbitrary nature, and the difficulties and ambiguities this sometimes causes. Why are there so many ways of spelling "their" there? Why should you have to twist your tongue if you want to sell sea shells by the seashore? And if you talk about a little girls' school, why should listeners be confused about whether the school or the girls are little? It may not surprise you to learn that the desire for a more well-defined and mathematically sound human language has been around for a long time. In fact, it has been over 50 years since Dr. James Cooke Brown first defined the language Loglan, a new human language based on the mathematical concepts of the predicate calculus. Later iterations of the language, after an internal political struggle against Dr. Brown by language enthusiasts, were renamed Lojban. In theory, Lojban, unlike English and other natural languages, is claimed to be minimal, regular, and unambiguous.

How do they define Lojban as such a clean language? First they made a careful choice of phonemes, basic sounds, chosen from among the ones most common in a variety of world languages. Each distinct-sounding phoneme is connected to uniquely defined symbols, removing any possible confusion about how to pronounce a given word: a word's sound is completely determined by how it is spelled. Then they defined a set of around 1,350 phonetically-spelled basic root words using these phonemes, being careful to not create homonyms or synonyms that could lead to confusion. The number of letters in a word and its consonant-vowel pattern determine what type of word it is: for example, a two-letter word with a consonant followed by a vowel is a simple operator, while five-letter words are what is known as "predicates". Replacing many aspects of parts of speech such as nouns and verbs from traditional languages, the formation of sentences is based around the predicates, which are in many ways analogous to the logic predicates of mathematics. For example, the predicate "tavla" means "$x1$ talks to $x2$ about $x3$ in language $x4$", with $x1$, $x2$, $x3$, and $x4$ being slots that may be filled by other Lojban words.

To get a better idea of how this works, let's look at a specific example. In the opening I alluded to the sentence "That's a little girls' school", which is ambiguous in English: is it a school for little girls, or a little school for girls? In Lojban, if it is the school that is little, the translation is "Ta cmalu nixli bo ckule". The predicate "cmalu" defines something being small. "Nixli" means "girl", and "ckule" means "school". The connector "bo" groups its two adjacent words together, just like enclosing them in parentheses in a mathematical equation, showing that we are talking about a school for girls, and it is that whole thing which is small. Alternatively, if we said "Ta cmalu bo nixli ckule", the virtual parentheses would be around "cmalu" for is-a-small and "nixli" for girl, showing that what is small is the girls, not the school. If there were no "bo" at all, there is a deterministic order-of-operations just like in a mathematical equation: the leftmost choice of words is always grouped together. So "Ta cmalu bo nixli ckule" and "Ta cmalu nixli ckule" are equivalent. Pretty simple, right? Well, maybe not, but after you stare at it for a while it kind of makes sense. And it does eliminate an ambiguity we have in English, at least for this case.

The adherents of these logical languages claim many potential benefits from learning them. They were originally developed to test the Sapir-Whorf Hypothesis, which claims that a person's primary language can determine how they think. Whatever the merits to this idea, I have a hard time seeing Lojban as a valid testing tool, unless a child is raised with this as his primary language without learning any natural languages – and that would be rather cruel to a child, I think! But many other virtues are claimed. Since the language is fully logical, it should facilitate precise engineering and technical specifications; a goal I can sympathize with, since I regularly deal at work with challenges of interpreting plain-English design specifications. It is also claimed as a building block towards Artificial Intelligence, since its logical nature should make it easier to teach to a computer than natural languages. It is also claimed as a culturally neutral international language, though it has fallen far short of other choices like Esperanto in popularity. And its adherents also enjoy it as a "linguistic toy", helping to research aspects of language in the course of building an artificial one.

At this point, I should add that I actually have a bit of a personal perspective on the viability of this kind of approach to technical specs. It is claimed that the logical and precise nature of Lojban will mean that if engineers would just learn it, all our specs would be clearer and unambiguous, leading to great increases in engineering productivity. But I work in the area of Formal Verification, where we are trying to verify chip designs, often having to convert plain-English specifications into logically precise formats. For many years there were different verification languages proposed for people to use in specifications, many offering minimal, highly logical, and well-defined semantics. But the ones that caught on the most in the engineering community and became de facto standards have not been the elegantly designed minimal ones, but the ones that were most flexible and added features corresponding more to the way humans think about the designs. So I'm a little skeptical of the idea that engineers would willingly replace English with a language like Lojban in order to gain more logical precision.

In any case, I think the biggest failure of Lojban has been that there are not enough people willing to learn it. Perhaps the human brain's language areas might just not be hard-wired in a way that naturally supports the predicate calculus. Even the lojban.org page states "At any given time, there are at least 50 to 100 active participants... A number of them can hold a real-time conversation in the language." So out of 50–100 people who are paying attention, only a subset of these can actually speak it? In comparison, Esperanto, an artificial international language designed by political idealists in the late 19th century, has tens of thousands of speakers, and an estimated thousand who learned the language natively from birth. And even Klingon, an artificial language invented for "Star Trek" and of no practical use to anybody, is rumored to have more fluent speakers than Lojban.

So, if you want to learn a cool way to think differently about language and make it more mathematically precise, go ahead and visit the Lojban institute online and start your lessons. But if you're hoping to make your engineering specifications more precise, communicate with your neighbors, or bring about world peace, you're out of luck. Remember to teach your children English as well.

When 'Is' Isn't
From Math Mutation podcast 196

Aside from the Lojban effort, others have attempted simpler solutions to improve the logic in natural languages. Followers of the cult-like philosophical movement known as General Semantics have found a way to supposedly improve the English language by transforming it into a new language called "E-prime". They base E-prime on the observation that poor and ambiguous usage of the verb "to be" causes many logical problems. This verb can hide the assumption of a logical hypothesis that needs better grounding in the user's previous definitions, and can create an illusion of a factual observation when in fact a writer merely states an opinion or guess. Thus, by completely disallowing usage of all forms of the verb "to be", we can speak more logically and consistently, and add clarity and rigor to our thought processes. This change to the language can also make statements and discussions less dogmatic, and some theorize that it could reduce strife and conflict in human societies. The name "E-prime" comes from the equation $E' = E - e$, where capital E represents the full English language, and lowercase e represents the verb "to be".

To get an idea of the problems E-prime attempts to solve, let's look at some example ambiguous or illogical statements in English that this change would disallow. Suppose you want to say "Erik is a podcaster". Do you mean that Erik makes his living at podcasting? That Erik records podcasts as a hobby? That Erik spends his weekends partying at a fraternity nicknamed "the podcasters"? You could mean any or none of these things. Another example comes from a classic Shakespeare quote, "To be or not to be, that is the question." We all know what that means, referring to Hamlet's struggles to decide whether or not to commit suicide. But we only know that due to lots of context from studying the play in school, or from hearing others talk about it. Wouldn't it improve the quote if Shakespeare had said, "To live or to die, I ask myself this." While you may dispute the poetic value of this rephrasing, I think you'll agree that it communicates the idea clearly, even to someone who doesn't have any context about Shakespeare or Hamlet.

This ambiguity in language comes from the fact that the verb "to be" can have many different meanings in English. It can signify identity, class membership, predication, existence, location, mathematical equality, and can function as an auxiliary verb. E prime advocates offer some common suggestions for making each of these types of statements clearer. For example, we can rewrite statements of title to include the exact credentials being asserted, as in replacing "he is the landlord" with "he owns the building and manages it." In statements about mathematics, using the precise term "equals" to replace "is" makes discussions much more logically sound, and helps to distinguish statements of equations from definitions of terms.

Some have claimed that while all these complaints of ambiguity are valid, eliminating a verb entirely from the language goes a bit too far, and that we should solve these issues by improving general discipline in language use. E-prime advocates tend to admit that this might make sense in theory, but in real life people just have too much difficulty applying that kind of discipline to their writing. For example, one web author writes, "After seven years of experience with this technique, I must agree with Dr. Kellogg (who even speaks in E-Prime) that, to work effectively, E-Prime requires the total elimination of be forms, since we use them addictively, even compulsively, as their subliminal residuum even in third drafts attests. On a recent foray into cyberspace, for instance, I found a Web Page featuring four sentences 'rewritten' in E-Prime – two of them containing be forms!"

Naturally, as we could expect with a movement for such a major change, some members of the General Semantics movement have opposed the idea of E prime as well. They point out that eliminating all forms of a particular verb can only make writing poorer and less interesting, by reducing a writer's options. They also point out that the forbidden verb communicates unambiguously in many common situations: for example, if someone asks "What color is that rose", the answer "That rose is red" contains no ambiguity or cause for confusion. They also point out that if we have the goal of eliminating logically unsupported inferences, malicious writers can always sneak them in using other verbs. They could change, for example, the non-E-prime "Erik's podcasts are silly" to the fully acceptable "Erik records silly podcasts", keeping the implied accusation of silliness without using the word "are". If you look at the article linked in the show notes, you'll also find numerous obscure objections to E-prime that have meaning mostly to General Semantics devotees.

So, should we all try to eliminate the verb "to be" from our writing, in order to make it logically clearer, more rigorous, and less dogmatic? Judging by what I see on the web, a small but dedicated E prime community seems to still exist, as a subset of the still-continuing General Semantics movement. I think the adoption of some aspects of their philosophy by the somewhat creepy Scientology movement didn't do them any favors. But as with many radical ideas, I think the concepts of E-prime may contain a kernel of truth. Several teachers of English and composition claim online that while they do not enforce the strictness of E-prime for general usage, they recommend it to their students as a way to improve discipline and clarity in their writing. After reading a few articles on it, I do feel motivated to try to review my writing for overuse of unsupported "is" statements. If you read carefully, you may already have noticed that I attempted to write this section in E-prime, with the only uses of the verb "to be" in cases where I am quoting it rather than using it. Do my thinking and speaking seem clearer than usual, or just more awkward? Only you readers can judge.

De-Abstracting Your Life
From Math Mutation podcast 212

One of the core principles of mathematics is the idea of abstraction, generalizing from various experiences to describe simplified models that enable rigorous reasoning. For example, if you look at a street map of your city, nothing there qualifies as a pure Euclidean triangle: all roads have thickness, varying slopes, squished raccoons, etc. But by reasoning about ideal triangles and lines, we can make powerful deductions about the distances between points that are very useful and accurate for practical purposes. However, there is a dark side to abstraction – when used too much in your daily life, it can cause you to over-generalize and lead to issues like stereotyping and prejudice.

For example, 20 years ago I had a Scottish roommate named Lloyd. Lloyd was a great guy, but I could not understand a word he said, due to his outrageous Scottish accent. Eventually we started keeping a notepad in the room so he could write down anything important he needed to communicate. After a few months with him, I was on the verge of insanity. Now, whenever I'm being introduced to someone from Scotland, I inwardly cringe, bracing myself for a similar experience. In effect, I have abstracted Lloyd as the general Scotsman in my mind, impacting my further relationships and experiences with his countrymen. It hasn't been that much of an impact in my life, as most residents of Scotland have yet to discover the joys of Hillsboro, Oregon, but it's still a bad habit. Is there something simple I can do to try to cure myself of this way of thinking? One intriguing set of ideas comes from a 20th century pop philosophy movement known as General Semantics.

General Semantics was first created by Polish count Alfred Korzybski in the 1930s, and detailed in a book called "Science and Sanity". This book describes a wide-ranging philosophy based on evaluating our total "semantic response" to reality, and learning to separate true reality, our observations of reality, and our language that describes the reality. By becoming conscious of our tendency to over-abstract, we can improve our own level of sanity, hence the book title "Science and Sanity". While serious philosophers and linguists generally don't consider Korzybski's ideas very deep, he attracted a devoted cult following, who believe that the General Semantics tools can significantly improve people's lives by reducing the errors that result from over-abstraction. This movement also led to the proposal for "E-Prime", the variant English language without the verb "to be", which I described in the previous section. Korzybski was also a bit of a math geek: when his Institute for General Semantics in Chicago was assigned the address 1232 East 56th Street, he had the address changed to 1234, in order to create a nice numerical progression.

Among the key tools that General Semantics provides for fixing over-abstractions are the "extensional devices", new ways to think about the world that help you to correct your natural tendencies. Many of these involve attaching numbers to words. The most basic is "indexing", mentally assigning numbers to help emphasize the differences between similar objects. For example, I might think of my friend Lloyd as Scotsman-1. Then, if introduced to another person from Scotland, I can think of him as Scotsman-2, emphasizing that he is a completely different person from Scotsman-1 despite their common origin. If I go out with my new friend for a yummy haggis dinner, I would think of the waiter as Scotsman-3, again recognizing his essential uniqueness and separating him from the other two. Through this assignment of numbers, I can avoid grouping them all into the single abstraction of Scotsmen, and help force myself to treat them as individuals.

Another important extensional device is called "Dating", similar to indexing but based on time. With this device, you attach dates to objects, indicating when you observed or experienced them. The Lloyd I remember should really be thought of as Lloyd-1993, since that's when I knew him, and I'm really only familiar with his characteristics at that time. If he emails me that he's coming to town, I should now think of him as Lloyd-2015, who may be a different person in many ways. Perhaps he has been working on his accent a bit, or maybe due to my 20+ additional years of engineering experience with colleagues from many diverse backgrounds, my ears have gotten better at discerning words in unusual accents. I should not over-abstract and assume that his most notable characteristics at one time, and my perception of them, will be the same today as in the past. Like everything in the universe, he and I are constantly changing, and I can use this extensional device to remind myself of that.

There are a number of additional extensional devices in General Semantics, such as the use of Et Cetera, quotes, and hyphens to further qualify your abstracted language. These seem a bit more awkward to me, though some may prefer them. Overall, I think the general concepts behind Korzybski's extensional devices probably can serve as a useful tool, especially if I go to Scotland sometime, though perhaps they are not quite as profound as General Semantics fanatics like to think. Korzybski's movement still seems to be going strong, with active institutes in New York, Australia, and Europe that have a presence on the web and in social media, and a quarterly newsletter still in print since 1943. Naturally, I've grossly oversimplified many of the core ideas for this short podcast, but if this has served to whet your appetite, you can find many other details at their website.

Your Kids Are Smarter Than You
From Math Mutation podcast 208

Did you know that, measured by constant standards, the average Intelligence Quotient, or IQ, of the world's population has been steadily increasing as long as it has been measured? In fact, by today's standards, your great-grandparents most likely would be formally diagnosed as mentally retarded. It's a little confusing, since the IQ tests are continually re-normalized, so the "average IQ" at any given time is pegged to 100. But if we look at the raw test scores and compare them across decades, we see that in every modern industrialized country, the IQ has slowly been creeping upwards. This effect is known as the Flynn Effect, named after the New Zealand psychiatrist who first noticed it in the 1980s. This seems pretty surprising – could our entire population really be steadily increasing its intelligence?

When I first heard about this effect, I was a bit skeptical. If you've read Stephen Jay Gould's classic *The Mismeasure of Man*, you have learned about all sorts of broken and ridiculous ways in which people have attempted to measure intelligence at various times. My favorite example was an IQ test from the early 20th century where your intelligence was, in part, dependent on your ability to recall the locations of certain Ivy League colleges. Even though such egregious examples no longer are likely to appear, you could easily hypothesize that the Flynn Effect was merely measuring the fact that over the past century, kids have been progressively exposed to a lot more miscellaneous trivia first through radio, then TV, growing mass media, and finally on the Internet.

Even simple things such as the expanding access to books and magazines throughout the 20th century might have contributed; I remember all the hours I spent biking between local used bookstores as a teenager, looking for cool math and science books, and I doubt my father had such an opportunity at his age. My daughter won't even have to think about such absurdities, having instant access to virtually all major literature published by the human race over the Internet. But it turns out that the belief that this IQ growth is just measuring access to accumulated factoids is not quite right – the growth has been very minor in tests dependent on this type of factual knowledge, and is really measuring an increased ability to do abstract reasoning using simple concepts.

In our modern lives, we take the concept of abstraction for granted: the ability to talk about and compare ideas, rather than just discuss concrete items and actions that are immediately relevant. And of course all of modern mathematics, including many topics we discuss on *Math Mutation*, is dependent on the ability to do this kind of abstraction. But this is not something to take for granted: it has been slowly growing in our society from generation to generation. For example, one article I found talks about a study done on an isolated tribe in Liberia. They took a bunch of random objects from the village and asked the villagers to sort them into categories. Instead of sorting into groups of clothing, tools, and food, as we might do, they put items together that were used together, such as a potato with a knife, since the knife is used to cut the potato. So apparently modern IQ tests are largely measuring our ability to think in abstract categories, and this is the ability that is increasing. Flynn has argued that we should really label this kind of thinking as "more modern" rather than "more intelligent" – can we really say objectively that one kind of thinking is better? However, we probably can say that this modern thinking is a critical component in the explosion of science and technology that we observe in the modern world.

There are numerous theories to try to explain the Flynn Effect. Most center on social or societal factors. Perhaps the explosion of media exposure is important not because of miscellaneous factoids, but because of the generally more cognitively complex environment, forcing us to think in abstractions to make sense of the massive bombardment of ideas coming at us from literature, television, and the Internet. The growth of intellectually demanding work, where more and more of us have jobs that involve at least some thinking rather than pure manual labor, may also contribute. Another possible factor is the reduced family size in the Western world: with fewer kids around, each gets more parental attention, and this may foster development of abstract thought. And of course, in recent years, I'm sure there has been an IQ explosion among the very important subset of the population who listen to *Math Mutation*.

Aside from social factors, there are more basic physical ones: essential improvements to health and welfare, such as massively reduced malnutrition and disease, could also be important here. Some studies have shown that simple health can have a much bigger effect on educational success than fancy computers. There is also the theory that we are simply measuring the effects of Darwinian natural selection, where parents with this more modern thinking style are more likely to reproduce, due to coping better in our technological 20th–21st century society. But most biologists believe that the Flynn effect has set upon us too quickly to be evolution-based.

To further complicate the discussion, some recent studies in Northern Europe seem to show that the Flynn Effect is disappearing or getting reversed. It's unclear whether this is a real effect, or an artifact of recent population shifts: over the past two decades, there has been massive immigration from the Third World into these countries, and it could be that we are just measuring the fact that a lot of new immigrants are in earlier stages of the Flynn Effect treadmill. But as in every generation, there is no shortage of commentators who can find good reasons why today's young whippersnappers are supposedly getting dumber, such as a focus on repetitive video games and social-network inanity. We need to contrast this with their parents' more intellectual pursuits, such as Looney Tunes and Jerry Springer.

So, what does this all mean? We certainly do see some effects in society that may very well be partially due to the Flynn Effect, such as the explosion of new technology in recent years. I think we should do whatever we can to continue making our kids smarter, and enabling more modern and abstract thinking – though of course, that would be true with or without the Flynn Effect anyway. Encourage your kids to engage in cognitively complex tasks such as reading lots of books, learning to play a musical instrument, and discussing cool math podcasts. But when they tell you in a few years that you're going senile, don't take it personally: you really are dumber than they are, due to the Flynn Effect.

My Brain Hurts
From Math Mutation podcast 56

I bet a lot of you out there, like me, often try to think of your brain as some kind of computer. It seems to make intuitive sense, at least, since it does consist of a bunch of neurons firing electrical patterns. In that case, our consciousness might simply be a set of mathematical patterns of electron firings, circulating around our computer-brains. But does this view actually make sense? Of course, many treatises have been written on the pluses and minuses of looking at the brain this way, not to mention the gods of numerous religions who would smite me just for making the suggestion. But recently I was reading an amusing little story called "The Story of a Brain" by Arnold Zuboff, part of the classic collection *The Mind's I*, that shows some interesting problems with this idea.

To summarize, the story centers on the brain of a man diagnosed with a terminal illness. Some scientists come up with a radical new procedure, where his brain will be preserved in a jar, and actually kept alive and conscious; stimulation will be provided by a computer that will fire the external nerves just as if the body were still attached. That way, the dying man could potentially live forever. So far, pretty standard science fiction stuff.

The twist comes when the scientists decide to try a new experiment. They separate the brain into two halves, and hook up each to a computer, designed to simulate the other half of the brain. They argue a bit about whether the halves have to be kept physically oriented the same way: can one be on its side? And do the computers have to be precisely in sync, or can they freeze the whole process for one brain-half during transport to another lab, then continue the program, effectively causing one half-brain to be on a time delay relative to the other? They eventually decide these issues should not be a problem, and happily carry off one half to another lab.

Years later, the experiment has been continuing, and now there are millions of labs all over the galaxy, each with one of the original neurons connected to a computer. Some of them are starting to wear out, so when this happens, whichever scientists owns that one simply acquires an equivalent neuron from another dying body, and hooks it up to the same computer program. Then one day, as he is bending over the neuron to observe it firing, one scientist has a thought. There must be some neuron among the millions in his head that is firing at each given moment when the lab neuron fires – so why does he even care about the one in the lab? As long as there are a bunch of people around with neurons firing continuously, the sum total of some subset is equivalent to the original subject's consciousness!

You can see the bizarre issues this story raises. If it is truly the case that the mind is a computer, and your consciousness is just a mathematical pattern going through it, where is the flaw in the story? When the brain was split, was the man's consciousness suddenly cloned, or did it cease to exist? What about when it was separated into neurons? Did the subject ever truly die, and if so, when? And why did the final scientist's observation not show that they could have just buried the guy to begin with, and not bothered with the lab?

It's fun to try to think of answers to these paradoxes. One obvious one is that there is a major premise: the computers that need to be hooked up to the brain-halves or neurons must effectively be artificial-intelligence simulators of the original subject's mind. Could such AI be effectively realized? Even if we could construct conscious software programs, an exact replica of an existing person's mind might be a tall order. Another issue is the final scientist's assumption that someone, somewhere, must be replicating a needed firing at every given moment. Given a finite human population engaged in a finite amount of neural activity, what are the true odds that this would happen? It's not clear to me the odds are anywhere close to 100%.

But these objections just scratch the surface; I'm sure that among the random patterns of neural firings in your brain, you can find many more ways to answer the points in this story. Some of you can probably find objections encoded in the random patterns in other parts of your body as well. But for deeper discussions of the many issues involved, I would highly recommend that you check out the classic anthology 'The Mind's I', where I first read the story.

Psychochronometry
From Math Mutation podcast 202

I recently celebrated my 46th birthday. It seems like the years are zipping by now – it feels like just yesterday when I was learning to podcast, and my 4th grader was the baby in the *Math Mutation* podcast logo. This actually ties in well with the fact that I've recently been reading "Thinking in Numbers", the latest book by Daniel Tammet. This book is actually a collection of loosely autobiographical essays about various mathematical topics. One I found especially interesting was the discussion of how our perceptions of time change as we age.

I think most of us believe that when we were young, time just seemed longer. The 365 days between one birthday and the next were an inconceivably vast stretch of time when you were 9 or 10, while at the age of 45, it does not seem nearly as long. Tammet points out that there is a pretty simple way to explain this using mathematics: when you are younger, any given amount of time simply represents a much larger proportion of your life. When you are 10, the next year you experience is equal to 10 % of your previous life, which is a pretty large chunk. At my age, the next year will only be 1/45th of my life, or about 2.2 %, which is much less noticeable. So it stands to reason that as we get older, each year will prove less and less significant. This observation did not actually originate with Tammet – it was first pointed out by 19th century philosopher Paul Janet, a professor at the Sorbonne in France.

Following up on the topic, I found a nice article online by an author named James Kenney. He mentions that there is a term for this analysis of why time seems to pass by at different rates, "Psychochronometry". Extending the concept of time being experienced proportionally, he points out that we should think of years like a musical scale: in music, every time we move up one octave in pitch, we are doubling the frequency. Similarly, we should think of our lives as divided into "octaves", with each octave being perceived as roughly the equivalent subjective time as the previous one. So the times from ages 1 to 2, 2 to 4, 4 to 8, 8 to 16, 16 to 32, and 32 to 64, are each an octave, experienced as roughly equivalent to the average human.

This outlook is a bit on the bleak side though: it makes me uneasy to reflect on the fact that, barring any truly extraordinary medical advances in the next decade or two, I'm already well into the second-to-last octave of my life. Am I really speeding down a highway to old age with my foot stuck on the accelerator, and time zipping by faster and faster? Is there anything I can do to make it feel like I have more time left? Fortunately, a little research on the web reveals that there are other theories of the passage of time, which offer a little more hope.

In particular, I like the "perceptual theory", the idea that our perception of time is in proportion to the amount of new things we have perceived during a time interval. When you are a child, nearly everything is new, and you are constantly learning about the world. As we reach adulthood, we tend to settle down and get into routines, and learning or experiencing something truly new becomes increasingly rare. According to this theory, the lack of new experiences is what makes time go by too quickly. And this means there *is* something you can do about it – if you feel like things are getting repetitive, try to arrange your life so that you continue to have new experiences.

There are many common ways to address this problem: travel, change your job, get married, have a child, or strive for the pinnacle of human achievement and start a podcast. If time or money are short, there are also simple ways to add new experiences without major changes in your life. My strong interest in imaginary and virtual worlds has been an endless source of mirth to my wife. I attend a weekly *Dungeons and Dragons* game, avidly follow the *Fables* graphic novels, exercise by jogging through random cities in *Wii Streets U*, and love exploring electronic realms within video games like *Skyrim* or *Assassins Creed*. You may argue that the unreality of these worlds makes them less of an "experience" than other things I could be doing – but I think it's hard to dispute the fact that these do add moments to my life that are fundamentally different from my day-to-day routine. One might argue that a better way to gain new experiences is to spend more time traveling and go to real places, but personally I would sacrifice 100 years of life if it meant I would never have to deal with airport security again, or have to spend 6 hours scrunched into an airplane seat designed for dwarven contortionists.

So, will my varied virtual experiences lengthen my perceived life, or am I ultimately doomed by Janet's math? Find me in 50 years, and maybe I'll have a good answer. Or maybe not – time will be passing too quickly by then for me to pay attention to silly questions.

CHAPTER 7

▪ ▪ ▪

Science and Skepticism

Ever since reading Martin Gardner's classic *Fads and Fallacies in the Name of Science* back in high school, I've had a strong interest in the idea of skepticism, critically examining our assumptions about science and the ways in which we misinterpret the world around us. This doesn't just mean picking at easy targets such as astrology and numerology, but also asking basic questions about the sources of all our scientific knowledge. We need to be careful not to define skepticism as simply our instinctive negative reaction to the sillier of the New Age fads, but as our requirement to carefully think about our hidden and not-so-hidden assumptions.

The "Skeptic Movement" has exploded in the past decade or so, aided by our improving ability to connect with like-minded folks on the web and to an abundance of excellent podcasts on the topic. In fact, I would guess that skeptic podcasts are today the single largest category of science podcasts on the web. Because of this overcrowding of the genre, I've tried not to focus too much on skeptical topics in *Math Mutation*. But every once in a while I can't resist producing an episode in this often surprising and amusing subject area.

Why Statisticians Stink at Statistics
From Math Mutation podcast 163

I've recently been reading the brilliant bestseller, *The Black Swan*, by Nassim Nicholas Taleb, which discusses the impact of improbable events in our lives. One of the surprising points Taleb makes is that humans are wired to be really bad at probability and statistics. We have a built-in heuristic mind, which takes shortcuts to jump to conclusions that would not be mathematically justified if we thought things through more carefully. Taleb makes many references to the work of pioneering researchers Kahneman and Tversky, who initially made this observation in the early 1970s, publishing classic papers on what came to be commonly known as "Cognitive Biases". Their experiments show that even professionals who should know better, like scientists, doctors, and statisticians, make these mistakes.

One common cognitive bias is called *anchoring*. Suppose you want to get an estimate of how many listeners *Math Mutation* has. Divide your friends into two groups, A and B. Ask group A, "Is the number of listeners more than 1000, or less than 1000?" Then ask them to guess the exact number. Afterwards, go to group B and ask

© Erik Seligman 2016
E. Seligman, *Math Mutation Classics*, DOI 10.1007/978-1-4842-1892-1_7

them, "Is the number of listeners more than 100,000, or less than 100,000?" Then ask them to guess the number as well. You will find that those in group A will guess much lower exact numbers than group B, with group A's guesses being close to 1000, while group B's guesses are closer to 100,000: apparently when asked for a number about which we know very little, we inherently "anchor" ourselves to numbers we recently heard as a reference point. Shrewd buyers and sellers make use of this bias, when they open a negotiation with a highball or lowball offer, anchoring the expectations in the directions they want. Surprisingly, some experiments show that the anchoring still takes effect even if the starting number is achieved by visibly spinning a roulette wheel.

Another common cognitive bias is called *framing*. Suppose I present you with this scenario: We need to figure out what podcast to play over municipal loudspeakers, to inspire the drug cartels to end their violence in a small Central American city, where 6000 people are killed per year. If you choose *Math Mutation*, projections show that 2000 lives will be saved. If instead you choose *A Gobbet Of Pus*, there is a 1/3 probability that 6000 lives will be saved, and 2/3 chance that none will. Which podcast will you choose? Most likely, your choice will be *Math Mutation*. But suppose we alternatively present it like this: if you choose *Math Mutation*, 4000 people will die. If you choose *A Gobbet Of Pus*, there is a 1/3 chance that nobody will die, and a 2/3 chance that 6000 will. In this case, the vast majority of people will make the other choice. But if you think carefully, you will realize it is exactly the same choice in both cases. The positive or negative framing of the question dramatically alters people's opinions of the result.

A third well-known cognitive bias is known as the *conjunction fallacy*. Suppose I give you a short bio of someone named Fred, who got 800 on his math SATs, studied physics at Princeton, and was champion of his chess team. Tell me which of the following statements is more likely: A. Fred is a bank teller. Or B: Fred is a bank teller and president of the local *Math Mutation* fan club. Chances are that you will gravitate to choice B. But think about this for a minute: choice B, being a bank teller AND fan club president, must be strictly less probable than A, which only says that he is a bank teller. We have taken the same basic statement, and added a more restrictive condition. You are fooled by your natural tendency to focus on specific statements in preference to more general ones.

So, what can we learn from all these cognitive biases? The real lesson is just to recognize that you have them: whenever you are dealing with numbers or probability, you will have a biological drive to take shortcuts rather than truly thinking the problem through. When you're just playing games or choosing podcasts, this may be fine – but when money or lives are at stake, you need to take that extra effort to step back and question the rationality of your decisions.

On Average, Things Are Average
From Math Mutation podcast 58

You may have heard of the "Sports Illustrated Cover Jinx", the legend that athletes who are featured on the cover of *Sports Illustrated* tend to suffer a decline in their careers afterwards. Naturally, we all like to think that divine powers are punishing these superstars for their excessive egos. After all, if the universe was fair, it would be *you* on that cover instead, right? But if you look at the careers of the cover athletes, there really may seem to be some kind of jinx effect: many athletes really do seem to achieve less after their cover appearances. So what's going on here? Actually, the explanation lies in simple mathematics.

The phenomenon influencing this situation is known as "regression towards the mean". This is a fancy way of stating that if you take any measurable quantity that has a sizable chance component, and you look at it again after you get a really high value, the next value you see will probably be lower. There's no magic here: *all* values have a higher chance of being close to average, assuming typical random distributions, so any given value you take will be most likely to be close to average. So if you get a really high value, the next one will almost certainly be lower – not because of any type of curse, but simply because on average, the value will be close to average.

What does this mean in terms of athletic performance? Well, I don't want to diminish the drive and effort of individual athletes, and it's certainly the case that superstars are usually very good at their games, with or without a *Sports Illustrated* cover. But I think we can all agree that there are numerous chance factors in any sports competition: the weather, the winds, the particular unpredictable strategies of the opposing team, the random fumbles that may or may not occur at opportune moments, the distracting yells from fans, or the intriguing mathematical podcasts from the previous day that are still distracting the athletes' thoughts. Together, these add up to a noticeable chance component in any athletic competition, and can add up to a sizable chance factor over a number of games.

So, if an athlete has an exceptional season that lands him on the cover of *Sports Illustrated*, that means that these random components of his performance all happened to line up to create a very high positive "plus factor" adding to the pure-talent part of his performance. Due to regression towards the mean, this plus factor is very unlikely to reach the same high positive level the following season: chances are that it will be close to average, as is the probability in any given season. Thus, while the athlete does not get any less talented, the random factors are probably not going to add up in his favor during the season after his *Sports Illustrated* cover. And it will look like the athlete is jinxed – but actually, all that has happened is that an exceptionally unlikely streak of luck has not repeated itself.

Regression towards the mean can pop up in many areas of your daily life, if you look for it. You probably won't have two car accidents for two days in a row – while you might credit this fact to consciously improving your driving after the first accident, it may just be that on average days, your bad driving isn't quite bad enough to cause an accident. Following an exceptional score on a test in school, you may be disappointed to find that your following tests are not so hot; maybe your studying slacked off, or maybe you just got lucky that one time. If you feel especially snotty one day and take a homeopathic remedy, you'll probably feel better the next day – not because homeopathic remedies do anything, but because you're likely to be close to your average health on any given day, unless truly suffering from a serious illness. Whenever you try to analyze some type of cause-effect relationship in your daily life, you should try to think about whether regression towards the mean might be the best explanation.

Don't Panic
From Math Mutation podcast 101

Suppose you arrive home after a long day of work, turn on your TV, and see a lurid report of a dangerous new disease, Exploding Brain Syndrome. Although it only affects about one out of every million people, its effects are horrific, and the story is accompanied by the usual footage showing what is left of "beautiful people" after their brains have exploded. Terrified, you go to your doctor to ask if you can be tested for this syndrome, just to be safe. "No problem," he says. "I have a test that is 99% accurate. Let me just take some blood, and I'll call you tomorrow." The next day there is a somber message on your voice mail. "Your test came out positive." Should you start preparing for the inevitable explosion of your brain?

Although most people, including many doctors, would say the answer is yes, this answer is actually very wrong. You need to take into account both the accuracy of the test AND the prevalence of the condition in the population, and once you combine those two pieces of data, you will see that you shouldn't be too worried. Let's put some concrete numbers on the discussion to illustrate.

Suppose you live in a city of one million people. Since the disease only affects one out of every million, there's probably only one person with Exploding Brain Syndrome in your city. But since the test is only 99% accurate, that means that of the almost one million healthy people, about 1% of them, or ten thousand people, will test positive on this test. So your actual chance of having the disease, given that you are one of the ten thousand and one people who tested positive, is one in ten thousand and one, or less than .01%. Not quite as good as the original one-in-a-million odds, but still a pretty minuscule thing to worry about.

This counterintuitive calculation actually has serious real-world implications: it means that over-testing for rare diseases is likely to result in huge amounts of time and resources wasted on treating false positives, and in the worst cases, actual health harm due to side effects of the drastic treatments necessary for some dangerous and rare diseases. For example, some studies suggest that PSA testing for prostate cancer may actually have done more harm than good in the years since it has become popular, with people who don't actually have the disease suffering from radiation and chemo therapies.

If your doctor tells you that you have a positive test for a rare condition, be sure to probe the details, get another independent test if available, and strongly consider the risks of treatment vs the risks of having the disease. There may be a stronger chance that your brain will explode from panic than from Exploding Brain Syndrome.

It Must Be True, There's an Equation
From Math Mutation podcast 132

It's said that the Russian empress Catherine the Great was annoyed by the atheist philosopher Diderot, and asked the famous mathematician Euler for help. So Euler came to her court and presented his proof, in the form of a mathematical equation:

"$(a+b^n)/n = x$, therefore God exists." Diderot, who had no training in algebra, could not answer this argument, and had to give up and return to Paris in embarrassment.

The story is actually an urban legend, but it still has somewhat of a ring of truth: equations do have an almost mystical power, convincing people that you have reason and logic behind whatever you're saying, regardless of the actual merits. I think all of us who took social science or philosophy classes in college had the experience of seeing the professor write an equation on the board, and wondering if there really was that level of mathematical precision in play. Even engineering professors are not immune to this bias: back when I was thinking of finishing my Ph.D. in computer science, I remember one professor critiquing my thesis proposal. "You need to get a few more equations in there." He didn't care what the equations were, or whether my main point truly required more equations: I just needed them there to demonstrate my scholarliness.

Physicist Alan Sokal is famous for blowing holes in academic pretensions. His most notorious stunt was getting a hoax article published that mentioned, among other things, the patriarchal social construction of the value of π. His classic book *Fashionable Nonsense* is full of examples of dubious references to math and physics concepts by postmodern academics. One example is his discussion of Jacques Lacan, the influential 20th century psychiatrist, whose writings often string many arbitrary-seeming math concepts together in order to sound impressive. Here's an example quote: "... The signifier can only be a line [trait] that is drawn from its circle without being able to be counted part of it. It can be symbolized by the inherence of a -1 in the whole set of signifiers. As such it is inexpressible, but its operation is not inexpressible, for it is that which is produced whenever a proper noun is spoken. Its statement equals its signification. Thus, by calculating the signification according to the algebraic method used here, namely S(signifier)/s(signified) = s(the statement), with S = -1, produces: $S = \sqrt{(-1)}$." Lacan goes on to explain that this is the imaginary number i as defined by mathematicians, and is somehow related to the human penis.

Those who misuse math like this have a lot of excuses. One major category is that the equations are mere illustrations or marginal elements, and that they should only be viewed in the full context of a work. But as Sokal explains, the misuse of these symbols and terms, which have very precise meanings that directly result in their usefulness in real math, shows a general disdain for facts, logic, and intellectual rigor, that *should* cast doubt on the entire context. Another major category of excuses is that we should allow some poetic license, as the math is merely as metaphor or analogy to relate abstruse concepts to something more familiar. But in nearly every case, the math is a lot less familiar to the typical reader in a humanities or social science discipline than the author's main argument: so exactly how is this use of math supposed to clarify anything? The logical conclusion is that in the vast majority of such cases, the math is brought in to add an impression of erudition and logical rigor to arguments that are otherwise weak or incomprehensible.

So, next time you read an article or book in a non-math subject area and see an equation, step back and think about it for a minute. Does it really make sense? Are all the terms used according to precise, sensible mathematical definitions? If not, you may merely be reading a piece of, as Alan Sokal puts it, "fashionable nonsense".

A Twisted Take on Turing
From Math Mutation podcast 107

Many physicists would agree that, had it not been for congestion control, the evaluation of web browsers might never have occurred. In fact, few hackers worldwide would disagree with the essential unification of voice-over-IP and public/private key pair. In order to solve this riddle, we confirm that SMPs can be made stochastic, cacheable, and interposable.

Did the previous paragraph make sense to you? Well, the sentences certainly sound plausible, in the endless flow of technobabble that seems to dominate today's online discussions. But in fact, they were randomly generated by a program called SCIgen, developed by a group of clever grad students at MIT. They form the abstract of a randomly generated paper called "Rooter: A Methodology for the Typical Unification of Access Points and Redundancy", which was actually accepted to the 2005 World Multiconference on Systemics, Cybernetics, and Informatics, a real conference despite its title, which sounds like it could be generated by the same program.

The program is built around a set of formal structures known as a "context-free grammar". Context-free grammars were first developed by Noam Chomsky in the 1950s to recursively describe basic language structures, decomposing chunks like sentences and phrases into simpler elements, until basic combinations of words are described. For example, a grammar that describes sentences where combinations of X and Y are added would have the following rule: S - > x | y | S + S. By starting with the single symbol S, and then choosing whether to substitute the terminals x or y, or the recursive addition of S + S, you can use this grammar to form results like "x + y", "x + y + y", "x + x + y + x", etc. Such grammars have quite a few uses in areas like parsing computer programs, or checking basic grammatical correctness of sentences. But an important thing to notice about these grammars is that they are context-free: you can't force any kind of real coherence with distant elements, which means they are inherently unable to represent the flow of intelligent thought that should be built in to a good academic paper. The fact that plausible academic papers can be built out of a context-free grammar of technical-sounding words is quite an embarrassment to academia.

Having spent time in the academic community as a grad student, and since then heard more than my share of low-quality conference presentations, I have to applaud Jeremy Stribling, Max Krohn, and Dan Aguayo, the authors of SCIGen. They correctly observed that too many academic conferences seem to have the sole purpose of making money for their organizers, and are latched onto by borderline academics who have nothing much to say, but want more publications on their resumes. I knew one professor who listed three dozen or so of his own papers as references in every new paper he published, and the titles all seemed like permutations of the same ten words. An excellent way to test the academic legitimacy of a conference' publications is to submit randomly generated papers, and see how many of them are accepted, in comparison to the acceptance rate for so-called serious papers.

In some ways, this is a variation on the famous Turing test, where an observer must converse with unseen entities in two remote rooms through online chat. The challenge is to distinguish the human from the computer program. Turing intended this to be a test of artificial intelligence – when the computer program could converse well enough to be indistinguishable from the human to the average observer, AI would truly be achieved.

This is kind of a reverse case though: rather than trying to prove artificial intelligence in their computer, Stribling, Krohn, and Agauyo were trying to expose natural stupidity in academia. I'm sure Turing would approve of this new application of his test though.

Their success at the conference I mentioned before was not a fluke. On the SCIgen website, you can see a list of conferences and journals that have accepted the randomly generated papers. Many of them have prestigious-sounding names, like the IEEE International Conference on Computer Science and Software Engineering, or the International Symposium of Interactive Media Design. Remember, just because someone forms smart-sounding words into sentences, it doesn't mean they are making sense. And if you see a lot of publications on a job applicant's resume, don't be too impressed, until you read, comprehend, and truly appreciate one of their papers. If it sounds too complicated for you to understand, it just might have been generated randomly.

I Want My Molecule
From Math Mutation podcast 7

Recently I was browsing in an otherwise respectable drugstore, and I was surprised to find some 'homeopathic' remedies available. As usual, I just shook my head and laughed. These so-called 'medicines' are created according to an alternative form of medical science that was created in Germany about two hundred years ago – and defies basic mathematics. Let's see why.

To start with, let's assume you accept the fact that substances are made out of molecules, as has been generally accepted for longer than my lifetime. Where the math fits in is that it's well known how many molecules of a substance are needed to make up its molecular mass in grams: approximately 6.02×10^{23}. So, as a simple example, if you have 10 grams of a medicine whose molecular weight is 10, you have about 6.02×10^{23} molecules.

Now we can figure out why the math of homeopathy just doesn't work. One of the basic principles of homeopathy is that a medicine does not lose its effectiveness, and in fact becomes more effective, after dilution. A typical step of dilution involves taking the original medicine and mixing it with one hundred times its volume of water, known as 1C. If you take a bottle of medicine, perform this '1C' dilution, then remove a bottle of the result, your new bottle has roughly 100 times fewer molecules of the medicine than you started with, or 10^{-2} times the original amount.

The fun part comes when you read homeopathic literature, and find that '30C' is a typical recommended dilution. This means that you go through this 1C dilution process 30 times. The first time you have 10^{-2} times as much medicine as you started with; the second gets you to 10^{-4}, and so on. By the end, you have 10^{-60} as much of the medicine as you started with. Multiply that by the 6.02×10^{23} molecules we started with in our example, and we are left with 6.02×10^{-37} – an infinitesimal fraction of a molecule. This means that we have a much bigger chance of winning the lottery than of having a single molecule of the actual medicine in a small bottle taken from the diluted sample.

I don't know about you, but when I pay good money for a medicine, I like to have at least one molecule of it in the bottle. Is that too much to ask?

The Gullible Ratio

From Math Mutation podcast 185

One potential podcast topic that bounced around in the back of my mind for many years is φ (phi), the golden ratio, the well-known irrational quantity equal to

$$\frac{1+\sqrt{5}}{2}$$

or approximately 1.618. The reason why I waited so long to create a podcast episode on the topic is that it's so overdone – the chances are that if you have even a passing interest in math, at some point you have read an article or been told by a teacher about the many amazing properties of this golden value and its connections to art, architecture, biology, and even human psychology. But did you know that, to a large extent, the concept of φ has been oversold, and is not nearly as significant as some claim? I came across this when reading an essay by Martin Gardner in his collection *Weird Water and Fuzzy Logic*, a set of articles debunking various pseudoscientific ideas; the chapter on φ is called "The Cult of the Golden Ratio." While φ does have some mathematical significance, its importance in other areas has been significantly exaggerated.

Before we start debunking it, let's review what φ is. Draw a line segment, and place a dividing point such that the ratio of the two segment lengths in the divided line is equal to the ratio of the length of the whole to the larger segment. Or, viewed algebraically, if the segments are of length A and B, we need $(A+B)/A$ equal to A/B. The ratio A/B is the definition of φ. The geometrical definition was first created by ancient Greeks and described in Euclid's *Elements*, where it naturally comes up in several theorems related to pentagons: the length of a pentagon's diagonal is equal to φ times its side. A related result is that in a pentagram, the star-shaped object formed from a pentagon's diagonals, each intersection splits its edges into golden-ratio segments. Another interesting geometric result, known as Odom's construction, is that if you inscribe an equilateral triangle in a circle, draw a line segment connecting the midpoints of two sides, and extend it at one end to intersect the circle, the segment's portions inside and outside the triangle are in a golden ratio.

These results seem to be a natural consequence of the type of ratio φ describes, but what is more surprising about this golden ratio are several non-geometric results. There are a couple of bizarre simple-but-infinite constructions that have been proven equal to φ. One is the infinite fraction

$$\varphi = [1;1,1,1,\ldots] = 1 + \cfrac{1}{1 + \cfrac{1}{1 + \cfrac{1}{1 + \ddots}}}$$

repeated out to infinity. That looks really complex at first, but if you define it as s, observing that s must satisfy the equation $s = 1 + 1/s$, the idea might be clearer. Starting from the latter form it's pretty easy to prove that s equals φ using the quadratic formula. Similarly, you can represent φ with the square-root-based construction

$$\varphi = \sqrt{1 + \sqrt{1 + \sqrt{1 + \sqrt{1 + \cdots}}}}.$$

But probably the most useful φ-related result is related to the Fibonacci sequence. As you may recall, the Fibonacci sequence is formed by continuously adding the previous two members of the sequence, so it starts out as 1, 1, 2, 3, 5, 8, 13, and so on. As you extend it out to infinity, the ratio between successive terms gets closer and closer to φ. Logically you would expect that this could result in some applications in nature, as there are several additive natural processes that can be modeled as a Fibonacci sequence, as previous cells or other components of an animal or plant add on to determine further development. However, since most such cases are finite in real life, it probably makes more sense to just describe them as applications of the Fibonacci sequence rather than uses of φ.

Where the pseudoscience comes in is when measurements are stretched or fudged to conform to the golden ratio based on very scant evidence. This is what has happened in most claims that the ratio is important in art, architecture, and nature. Some have taken this concept to ridiculous extremes – in a search for φ-related books on Amazon, I even came across one called *The Golden Ratio Lifestyle Diet*. Here's part of its description: "The Diet is based on a fascinating formula – the Golden Ratio – that when applied to anything, creates greater beauty, unity, efficiency, value and success – a greater whole exceeding the sum of its parts. This formula guides the structure of matter and movement of energy throughout the Universe." Hmmm... I think I'll stick with the much yummier 'pi diet' myself.

One common misconception that is the source of much of this insanity is that somehow elements in this ratio are provably most aesthetically pleasing to the human eye, and thus φ is wired into our biology. This largely stems from a 19th century study by Gustav Fechner, where he showed a bunch of rectangles to subjects and asked them to choose the prettiest. But the many subsequent attempts to replicate this study have gotten inconsistent results. In fact, Gardner describes one replication attempt that came to the conclusion that 3×5 index-card-like rectangles are equally preferred, with no irrational values needed – perhaps this is a case where the free market has optimized better than the mathematicians.

There have also been many observed measurements of this ratio in ancient architecture, such as claims about Egyptian pyramids and the Greek Parthenon, and in various body proportions of creatures and plants. All these suffer from a fatal flaw: the lack of precision. As Gardner points out, by slightly altering the starting and ending positions of a measurement, any measurement that is somewhere close to a ratio of 1.6:1 or 3:5 can be tweaked to seem to match φ or its inverse to a few significant digits. And there is so much individual variation in nature that it's nearly impossible to create firm rules about any bodily ratio. For example, a 1992 study checked the navel heights of 319 men and women, to check claims that the navel divides the body in a golden proportion in humans. Their conclusion was as follows: "Admittedly, measurement error in this study was a ticklish problem. But... a practitioner of the navel art who wishes to capture divine proportions would have to place the belly-button of a typical Middlebury College coed about 0.7 inches higher than it really is."

So, is the Golden Ratio important? I think there is no doubt that it is a quantity of mathematical significance, due to its intimate connection with geometry and the Fibonacci sequence, as well as its appearance in some bizarre infinite constructions. But we need to be careful before claiming a particular irrational number as somehow psychologically pleasing, built into biological organisms, or possessing some kind of mystical power. I think online forum participant 'MartianInvader' has a good summary: "Reports of the golden ratio's awesomeness are often overstated, but there's still a decent amount of awesome in there."

Monkeying Around with Probability
From Math Mutation podcast 118

Recently in creationist circles there has been an argument making the rounds that simple probability shows that evolution cannot occur, based on calculations of the odds of mankind evolving vs the amount of matter and time in our universe. When I first read one of these arguments, it seemed like they might have a point; but after thinking about it a little while watching the monkey-like behavior of my daughter, I realized that there are at least three major flaws. I'll start by describing the argument – see if you can spot the problems before I tell you.

One amusing form of the argument is found at a *Free Republic* posting by someone named Brett Watson. He tries to calculate the probability of monkeys randomly typing "To be or not to be, that is the question"; we would probably agree that any simple life form is at least as complicated as this sentence. Assuming we have a basic 32-key keyboard, the probability of a monkey typing every one of the 41 characters in this sentence correctly is 1 in 32^{41}: 1/32 chance of the first character being right, $1/32^2$ chance of the first two characters being right, and so on. Assuming the monkey types one line per second, a few simple calculations show that even if typing for 17 billion years, the monkey will have less than a 1 in 10^{39} chance of succeeding. And if you allow 17 billion galaxies, each with 17 billion planets containing 17 billion typing monkeys, there is still less than a 1 in 10^{13} chance of success. So the conclusion is that life in general could never arise by chance.

The first major flaw is perhaps the simplest. Sure, there might be only a 1 in 10^{39} chance of a particular universe evolving life – but why is it so inconceivable that the number of planets in the universe or in the multiverse is that large? As we've mentioned in an earlier chapter, we have no way of knowing how much exists beyond the frontier of space that we can observe. In addition, the 'many worlds' interpretation of quantum mechanics seems to imply exponential numbers of alternate universes branching off at every moment. And even if that's wrong, and the number of universes is fixed and finite, we really have no way of knowing how many parallel universes there are. If we admit the possibility of a large unobserved portion of our universe, or of parallel universes, why should 10^{39} planets be less conceivable than a handful? The well-known *anthropic principle* states that we are observing one of the lucky few planets/universes that did evolve life, simply because otherwise we wouldn't be here to observe it.

If that doesn't satisfy you, or you still find the idea of parallel universes perplexing in general, a 2006 article by John Allen Paulos points out a second flaw. Look at the forms of life that we are familiar with. Do we really think these are the *only* types of life that could have evolved? Perhaps there are many alternate ways to build life from chemicals in our universe, just as viable and self-reproducing, that could have evolved instead; we just happen to observe that ones that were lucky enough to evolve on our planet. Looking at the monkeys-typing-Shakespeare analogy: Do they really have to type "To be or not to be, that is the question"? Maybe they just had to type *any* valid English sentence, which is much easier. Paulos uses the analogy of a deck of cards: suppose you deal out the 52 cards in order. There are about 10^{68} possible orderings: but you are guaranteed to get some ordering when you deal the cards. No matter what order the cards were dealt in, you can point and say, "Wow, there was only a 1 in 10 to the 68th chance of that ordering – divine beings must have intervened!" We have no way of knowing exactly how many ways life can form, but even within the limited parameters of life on our planet, scientists are continually surprised by discoveries of life previously thought impossible in places like Antarctic ice or geothermal vents.

A third flaw in the monkeys argument is a bit more subtle, but is perhaps the most powerful. It totally ignores the concept of "survival of the fittest" in evolution. In other words, evolution is a feedback system, where the forms that are able to reproduce succeed in propagating, while the non-viable forms die out. This even extends back into precursors of life, organic molecules that can reproduce themselves or catalyze other organic reactions. Thus each event in the evolution of life is not independent: when early life forms or their precursors get a partial success, this success is built upon in further evolution. This is critical, since the calculations of infinitesimal odds are based on multiplying many small probabilities, which is only a valid operation if the events are independent. Back to the typing monkeys analogy, not only are the monkeys allowed to produce any valid English sentence, but there is a guy with a whip standing behind each monkey, thrashing him if he starts to type a character that will not form a sentence. In this scenario, I bet we would start getting valid sentences pretty quickly. Assuming the monkeys did not revolt, slay their cruel masters, and form their own non-Shakespeare-typing society.

So, does all this prove that evolution took place, or that creationists are wrong? I'm afraid that discussion is beyond the scope of this book, though I think you can find about 10000 podcasts on each side of this topic if you're interested, not to mention uncountable numbers of online blog-comment flame wars. It's also possible that future developments in biology and physics will invalidate the counter-arguments I presented here: maybe it will be conclusively shown that there are only one or a handful of universes in existence, there is only one way life can evolve, or that there is an unbridgeable chemical gap between the primordial soup and true precursors of life. But I think you can see from these points that today's simple probabilistic objections to evolution are missing some important factors. If you want to argue for creationism over evolution based on today's knowledge, you'll have to look elsewhere besides in the math.

Solving Burma's Problems
From Math Mutation podcast 42

Reading the many articles about current events in Burma, I noticed that back in 1987, the leader Ne Win decided to replace the currency with notes in units of 45 and 90, since his lucky number was 9. But that led to mass protests and violence, perhaps due to its sanity- challenging interpretation of economics and politics, and helped contribute to the opposition movement that continues to this day. However, if we assume for the moment that our destinies are controlled by capricious divine beings who examine arbitrary numbers that come up in our lives and choose whether to help us based on their correlation with our selected lucky numbers, I think the general still got it wrong.

While harnessing the powers of the number 9, there are a few other aspects that the general should have taken advantage of as well. You may have heard of "casting out nines", a method of checking arithmetic calculations by adding up the digits of all the numbers, finding the remainder modulo nine, and checking that the calculation using the remainders matches the original numbers. It couldn't hurt to mandate that everyone use that – what could be luckier than double-checking your arithmetic? And then, there's the fact that six recurring nines appear in decimal places 762 through 767 of π. So, in any calculation using π, it should be required that you calculate it to 767 digits of accuracy, to maximize this luck factor. That might be a bit of a pain, but after all, what's a little pain when compared to the extra luck that comes from using such auspicious numbers? Finally, since 9 is the atomic number of fluorine, we must be sure to fluoridate the nation's water supply. Having had recent root canal work, I'll always consider healthy teeth especially lucky.

But there is one fundamental flaw with all these properties, except the chemistry of fluorine – they all depend on the fact that we are using the ordinary base 10 representation of numbers. For example, if 9 is your lucky number, why is 90 a good currency unit? Sure, 9 is a factor, but then it's multiplied by 10, which may or may not be particularly lucky. As best I can figure, somehow ending in a 0 adds luckiness to a number, according to Win's theory. But if that's the case, why are we restricted to base 10 representations? In fact, with so much modern economic activity (even back in 1987) passing through computers, I think the binary representations used within computer systems are much more relevant. While 90 in binary does end in a 0, since it is 1011010, we can do much better. Any number divisible by 8 ends in *3* zeroes in binary, since 8 is 2 to the 3rd power. So, if we take 9 times 8, or 72, this is 1001000 in binary, making it three times as lucky as 90. In other words, due to its superior binary luckiness, Ne Win should have chosen 72 as a major currency denomination.

But let's take a step further back. Our choice of 72 or 90 is based on something rather superficial, how many 0s appear at the end of the number in various representations. If we assume that we can find a numerologist who attended math classes beyond third grade, they probably learned that numbers also have deeper properties that are independent of a particular representation. How about taking the various powers of a number? 9 is especially auspicious for this purpose: being a perfect square, we can take both integral powers and half-powers of 9 and get nice clean integers for our currency denominations. 9 to the 1.5 power is 27, 9 squared is 81, and 9 to the 2.5 power is 243. So if we choose 9, 27, 81, and 243 for our currency denominations, we are invoking the super extra lucky powers of the number 9, independent of any particular chosen representation.

There is one inescapable conclusion to draw from all this: the source of all Burma's problems for the past two decades is obviously the fact that they are using incompetent numerologists. All they have to do is hire me, and due to my superior understanding of the number 9, I'll get the country back on track.

A New Numerology
From Math Mutation podcast 128

I was amused recently to read an article on "numerology", the New Age technique where your personality is analyzed based on aspects of your name or birthdate. A typical numerological system will work like this: get the number associated with each letter of your name, then add the numbers together, and then keep adding digits together until you have a total that is one digit. So for example, in my name 'Erik', E is the 5th letter of the alphabet, R is the 18th, I is the 9th, and K is the 11th. $5 + 18 + 9 + 11$ is 43, and $4 + 3$ is 7, so I am a '7' personality. I look up that number on a numerological table, and find that it is associated with thoughtfulness. So I guess I must be thoughtful.

Being thoughtful though, I have to wonder why there should be a mere 9 personality types that describe everyone. Of course various numerology systems add additional wrinkles, such as 'master numbers' 11, 22, 33, and 44 that should not be reduced, but still, it seems like something more is needed. Why not treat each individual letter position, for example, as having its own meaning? That's how we enable multi-digit numbers to signify larger values: for example, 123 is 3 times 10 to the 0th power, + 2 times 10 to the 1st, + 1 times 10 to the 2nd. That's how with only three digits from 0 to 9, we can describe 1000 different numbers. If we just added the 1, 2, and 3 together, we couldn't tell 123 from 321, and both would be equivalent to 42 or 6. Shouldn't there be thousands of personality types, at the very least? If the New Age gods are somehow using our names to encode important information about our lives, I would think they could take advantage of the same elementary principles in our place-based number systems. Hopefully most of these gods have the 6th-grade education necessary to understand such things.

While we're generating numerical values to describe our personalities though, does it really make sense to base them on our names? A name is something that can easily be changed, and is often ambiguous if translated from another language. How about getting something physical from our bodies to generate a large number, each of whose places will be significant? Wouldn't it be great if there were some consistent aspect of our body, such as something inside our cells, that we could use to extract a descriptive number?

Actually, it seems that biologists have beat us to it. Inside each of our cells are long strands of DNA, which are basically strings of 3 billion or so "base pairs", each consisting of one of four basic nucleobases and its complement. And as we all know, these genes determine many aspects of our bodies and our personalities. Thus we can think of our DNA sequences as extremely long base-4 numbers which describe us in minute detail. Genes have been identified with such varied characteristics like eye color, vulnerability to cancer, and introverted vs extroverted personalities. So, ironically, the numerologists were right all along – each of us does have a unique number that describes us in detail, and can be used to help understand our personality and our health. It's a little more complicated than the ones they usually use, but I'm sure they won't object to correcting their systems in light of the latest science, right?

Somehow I'm not holding my breath waiting for online numerology sites to correct their name-based systems and add links to genetic analysis labs. But who knows, perhaps all this DNA stuff is just a trick by the dark sprits to distract me from truly understanding that I am fundamentally a Number 7 guy, and using the power of my number to drive the demons from the earth.

One Intestinal Worm Per Child
From Math Mutation podcast 110

Most of us have seen, in one form or another, heartbreaking pictures of deprived children in the developing world. Many of them live in huts that seem to be made of discarded particle board or scrap metal, with floors that consist, like the road outside, largely of mud. Often the children are seen playing in the streets, with little but dirt and garbage as their toys. We see obvious signs of disease and starvation, such as distended stomachs. And there appears to be little sign of hope or opportunity for the children growing up in these environments, without some fundamental changes to their way of life. So, as enlightened Westerners, our duty is clear. We must ensure that each of these children has an inexpensive laptop computer.

What, you're saying that's not your reaction? Well, before going on to discussing the details, let's talk about a basic mathematical principle involved here. Amdahl's law, as stated by computer architect Gene Amdahl in 1967, points out that if you have a process that can be broken down into many subprocesses, your overall speedup from some improvement is limited by which subprocesses you can do more efficiently. Suppose we have subprocesses $P1$ and $P2$ that take up time $T1$ and $T2$, and you can't improve $P1$, but can make $P2$ run n times faster. Then the total runtime is $T1 + (T2/n)$. So if $T1$ is much larger than $T2$, then even with a huge value of n, representing a huge improvement to process $P2$, you still have to run the full time of $P1$.

To make this more concrete, suppose you have a computer program that reads a list of names, then sends an email to each person telling them to listen to *Math Mutation*. Suppose reading the file takes 90 seconds, and sending the email takes 10 seconds. Then if you get a new multicore computer that can send email at lightning speed, 1000 times faster than before, but doesn't read files any faster, what's our overall improvement? Well, assuming the time for sending email is now negligible, the total time went from 100 seconds to 90 seconds, for a speedup of about 100/90, around 1.1. It doesn't quite match the 1000 times speedup claimed by the vendor, because there was a huge part of our process that was not improved.

How does this relate to solving poverty through laptop computing? I have to admit I'm abusing Amdahl's Law slightly here, since improving education isn't quite the well-defined process of a typical parallel computation. But for the moment let's accept that a primary problem in these third world areas is educational opportunity, and there are many factors that affect educational quality. Let's also assume that our goal is to, for the majority of children, reach some basic educational baseline where they achieve respectable scores on international tests of math and literacy.

I'm sure there are many factors that affect test scores in such societies. One of them might be lack of access to computers. But are there other factors that have more importance? Again, computer access is certainly nice, but must be considered alongside other possible issues. And then, following Amdahl's Law, the largest factors are the ones that charitable donors should concentrate on, as improvements in other areas will have an inherently limited effect. These issues came to mind as I was reading a recent Miller-McCune article on experiments comparing the One Laptop Per Child concept with two other initiatives in rural India. One involved providing funding for de-worming, on the assumption that a large proportion of school nonattendance was due to children battling intestinal worms. Another program decided to address teacher absenteeism, a huge problem in remote areas: they provided digital cameras to teachers, and required that teachers send in a date-stamped photo with their students each day in order to get paid. Both of these programs showed much more significant gains in student achievement than providing computers. Apparently when students are too sick to make it to school or their teacher doesn't show up, that hurts education a bit more than lack of technology.

What's the lesson here? When looking at any sort of problem, in areas of public policy as well as engineering, we really need to keep Amdahl's Law in mind. For maximum effectiveness, we need to always look at the multiple factors and try to estimate how much of the overall problem they relate to, rather than choosing a particular initiative tied in with our personal emotions and hobbies and make that the focus. When we look closely at what is really needed, the answers may surprise us.

I don't want to get too cynical about the 'sexy' initiatives proposed by high-tech wizards to improve the third world: their hearts are surely in the right place. Perhaps they even have done some studies that disagree with the ones I read about, and really do believe lack of computers is the number one factor holding back Third World education. But I haven't seen much evidence of this in the many media articles on the "One Laptop Per Child" initiative. I think Amdahl's Law and similar factors, rather than simple emotional appeals, need to become a key component of all discussions of these kinds of initiatives. Journalists need to do a better job of trying to look at the various contributing factors, and show estimated weights in terms of effect on the overall problem, in order to truly examine the best ways to direct charitable efforts.

CHAPTER 8

■ ■ ■

Analyzing the Arts

There has always been an intimate connection between mathematics and the arts. Many of the math geeks I've known have had a strong interest in various forms of art. Back in college, I remember the majority of my fellow math majors being skilled with some kind of musical instrument. Rudy Rucker, one of the brilliant popular-math authors who originally inspired *Math Mutation*, is also known as a literary pioneer in the "cyberpunk" movement, as well as being an amateur painter in his spare time. Another inspiration was Douglas Hofstadter's groundbreaking classic *Godel, Escher, Bach*, which explored connections among math, art, and music. And as we discussed earlier, Edwin Abbot introduced the idea of the fourth dimension through a literary allegory rather than through dry exposition.

Perhaps the most obvious area of connection between math and art is in the visual arts and their close connection to geometry: the core idea of perspective, that we can get a real 3-dimensional impression by looking at a simple 2-D surface, has always amazed me. Music is highly mathematical as well, as even the ancient Greeks recognized the connection between musical tones and mathematical ratios. While it might not be quite as obvious, our study of literature can also benefit from mathematical insights. The articles in this chapter should help open your mind to the many connections between art, music, literature, and mathematics.

Discovering the Third Dimension
From Math Mutation podcast 135

You may recall that a few chapters ago, we made fun of some popular singers who changed a lyric about discovering the fourth dimension to one about discovering the third. This got me thinking about when the 'third dimension' might have actually been a new discovery. In a literal sense, this is a silly question, since a human who wasn't aware of the third dimension would never be able to get out of bed in the morning to ask it. But on the other hand, there are some areas in which the third dimension was discovered rather late in human history: in particular, in art. We've all seen those flat ancient Egyptian pictures on the pyramids, with those wacky arm angles due to the total inability to represent depth. Then at some point, artists began to realize that by paying a little more attention to how things should look, they could produce a realistic perspective view of the third dimension. When did this happen?

© Erik Seligman 2016
E. Seligman, *Math Mutation Classics*, DOI 10.1007/978-1-4842-1892-1_8

To start with, let's review the basics of perspective in art. I think artists have always intuitively known that closer objects should appear larger than ones that are far away – but this alone is not sufficient to create an illusion of three dimensions. To create this illusion, an artist needs to conceptualize a horizon line, representing the distant line where the sky meets the ground; and one or more vanishing points, where parallel lines eventually meet in the distance. Effectively, the artist is imagining the canvas as a window into another world, thinking of the paths of the rays of light from the depicted objects to the viewer's eye, and painting the images where these rays would hit the canvas. As usual, this barely scratches the surface; you can find a lot more info online.

When I first thought about this topic, I assumed I would be looking up developments in the early Renaissance, based on my vague recollections from high school classes. But I was surprised to find that actually, perspective in art was discovered by the ancient Greeks, in the 5th century BC. The Greek version of this technique seems to have originated with a scene-painter named Agatharcus, who was trying to create realistic backgrounds for the plays of Aeschylus. Some sources do also say that there were some occurrences of perspective in painted vases from the previous century, so it may not have been entirely original even then. In any case, Agatharcus was apparently very talented for his time, and in heavy demand – according to one anecdote, the prominent general Alcibiades held him hostage in his house for three months and ordered him to paint it. Perhaps Alcibiades had had a bad prior experience with a contractor who walked off the job; this is one aspect of human society that has probably remained constant for several thousand years.

The realistic illusion of three dimensions in a stage background was striking enough to attract the interest of leading mathematicians and philosophers of the day. According to the Roman historian Vitruvius, Agatharcus' paintings "led Democritus and Anaxagoras to write on the same subject, showing how, given a centre in a definite place, the lines should naturally correspond with due regard to the point of sight and the divergence of the visual rays, so that by this deception a faithful representation of the appearance of buildings might be given in painted scenery, and so that, though all is drawn on a vertical flat facade, some parts may seem to be withdrawing into the background, and others to be standing out in front."

Sadly, we don't have any surviving examples of Agatharcus' painting, and none of the related written works survive, though Vitruvius's description seems clear enough to leave little doubt about the Greek discovery. In addition, we do have solid visual evidence that the ancients did know about this technique: in the ruins of Pompeii, from the first century AD, a paintings were found that clearly show the use of perspective. These were probably copied from Greek originals. Looking at them, you are likely to be surprised that they came out of the ancient world.

Figure 8-1. *Pompeii fresco showing perspective[1]*

So, their modified lyrics to "As Time Goes By" could make sense, if Binnie Hale and Rudy Valee were born around the 5th century BC. On the other hand, we should probably have learned by now that it's foolish for us to try to make sense of any pop music lyrics.

---

[1]Sourced from Wikimedia Commons at https://commons.m.wikimedia.org/wiki/ File:Pompeii_Fresco_001.jpg, public domain under tag {{PD-Art}}.

A New Perspective on Perspective
From Math Mutation podcast 154

I recently came across a passing reference to a statement by 20th century modern art theorist Marshall McLuhan, claiming that visual perspective is not an inherent aspect of art, but a cultural artifact of our society. In other words, the lack of perspective in primitive art is not some kind of lack of sophistication, but a result of a different cultural worldview and choices, which have not taught them to see the world in a way in which perspective would make sense. To a member of a culture that did not use this technique, perspective-based art would be meaningless until they were taught to understand it as a representation of reality. Finding this a bit insane, I decided to do some web searching to see if people could really believe this. Are realistic depictions of perspective just something we learn to expect in pictures, rather than a fundamental step in artistic maturity?

As we discussed in the previous section, artistic perspective is a set of methods for realistically depicting the 3-D world on a two-dimensional canvas. It makes use of concepts like a horizon line and vanishing point to create the illusion of viewing a scene as if the viewer were really present in the landscape. Aspects of perspective were visible in Western art as early as the 5th century BC, though a systematic approach based on an understanding of geometry did not begin until the Renaissance. The fact that artistic perspective has a close relationship to culture is beyond dispute; the objective view of reality that leads to perspective was also the underpinning of the scientific worldview that enabled the evolution of modern science.

We should note that these techniques are not perfect: in the real world, for example, the true horizon line is curved due to the curve of the Earth. We also have the naturally curved shape of our eyeballs to contend with, creating slight changes in how visual images arrive. Furthermore, no non-holographic painting can truly reproduce the 3-D view of the world we get from our two eyes viewing it at once. It is also true that some cultures have simply believed in different purposes for art; for example, in Egyptian and Byzantine works, figures are often placed and sized according to their relative importance in society, rather than attempting any kind of realistic view. But I don't think you can deny that Renaissance art was attempting a concrete task, realistic depiction of the world, in a way that these other kinds of art simply are not, and achieving it at a much more advanced level.

McLuhan and his colleagues seem to have more fundamental concerns. Complaining about the limitations of artistic perspective, they make statements like "to close one eye and hold the head still at a single predetermined point in space is not the normal way of looking at things". But they go much further. McLuhan once wrote, "far from being a normal mode of human vision, three-dimensional perspective is a conventionally acquired mode of seeing, as much acquired as is the means of recognizing the letters of the alphabet, or of following chronological narrative." They view perspective as a kind of pictorial code that we have learned to understand in our particular culture. Well-known art critic Herbert Read added that perspective "is merely one way of describing space and has no absolute validity".

Their biggest concern seems to be that perspective-based art separates the viewer from the object being viewed. Art critic Robert Romanyshin writes, "The painter (and the viewer) imagines that he or she is looking at the subject to be painted (the world to be viewed) as if through a window... The condition of the window implies a boundary between the perceiver and the perceived. It establishes as a condition for perception a formal separation between a subject who sees the world and the world that is seen; and in doing so it sets the stage, as it were, for that retreat or withdrawal of the self from the world which characterizes the dawn of the modern age."

And I think this is ultimately what McLuhan and his colleagues are interested in: they are trying to utilize this odd philosophy of art in the service of a new, more holistic, world view. This seems well-aligned with the modern New Age movement, with their many complaints that objective science is missing something, and we need to view ourselves as part of nature and reality rather than trying to logically observe it from outside. It goes hand-in-hand with the tendency to idealize primitive cultures and reject scientific progress. While many postmodernists find this an emotionally satisfying view, and refuting it is way beyond the scope of this book, I find it rather unconvincing. I don't see how thousands of pages of philosophic discussion can undo the simple act of placing a primitive painting, a Renaissance painting, and a photograph side by side.

But it is fun to do a little thought experiment for a moment, and accept the proposition that interpreting perspective-based art as a 3-D world is merely something we have learned in our culture. Could it be that there is some clever way to depict FOUR dimensional figures on a canvas that our culture simply has not picked up on yet? Maybe our lack of ability to visualize four-dimensional objects is just a consequence of our current cultural limitations, and not a fundamental limitation of our 3-D biology. At some proper cultural moment, a grad student at the Pratt Institute will suddenly paint a picture that will make the fourth dimension perfectly clear and logical to us. From that point, we will easily be able to visualize four-dimensional geometry, and laugh at our primitive forbears whose visual understanding was limited to three dimensions. Four-dimensional illustrations will become commonplace, and after work we will relax by playing 4-D adventure games on our Wii 4D game systems. At that point, I'll owe McLuhan an apology.

Hippasus's Revenge
From Math Mutation podcast 142

The relationship between mathematics and music was discovered over 2000 years ago, by the ancient Greek mathematician Pythagoras. Legend has it that he was wandering by a blacksmith's shop one day, and noticed that the clanging of the hammers against the anvils had a musical quality to it, with the notes depending on the weight of the hammers, and that certain pairs of hammers sounded better together, if their weights were in whole number ratios. Numerous authors have pointed out that this story is pretty unlikely: without careful hammer and anvil measurement and construction, which a blacksmith would be unlikely to do, it's nearly impossible that a blacksmith's hammers would harmonize well. It's much more likely that this observation came from experimenting with the strings on Pythagoras's lyre, an ancient Greek stringed instrument.

Pythagoras observed that if he vibrated strings whose lengths were whole number ratios, they would sound good together. And if two strings were in a 2:1 ratio, they would sound especially harmonious, being in some sense the "same note" in different octaves. Why would this happen?

Well, remember to think about how sound travels through the air: you've probably seen drawings of sound waves, with peaks and troughs at regular intervals, which are interpreted by the human ear. If two waves have frequencies that are whole number ratios, they will periodically sync up, with some peaks occurring together. For example, suppose you play an A note, at 110 Hz, and then play a higher A from the next octave at 220 Hz. If you think about it, and draw some squiggly lines, you will see that each wave peak of the lower note will correspond to a peak of the higher note. To a listener, this creates a pleasant effect, and the notes sound good together. Other ratios like 3 to 2, 5 to 4, etc, also create effects that sound good together as the peaks and troughs of the waves periodically line up, forming various chords and other musical structures. The system now known as Pythagorean Tuning was based on using frequencies in a 3 to 2 ratio, which sounds very pleasing to our ears.

Once they discovered these relationships, the Pythagoreans took the idea to extremes that sound a bit odd today. They looked at the apparent distances from the known planets to the Earth and decided that these ratios must form a "music of the spheres", a cosmic symphony played by the gods. Looking at the ability of these mathematical ratios to invoke various emotions, they also believed that music would have the power of healing. At the major medical centers of ancient Greece, treatment would always be accompanied by appropriate music. Roman philosopher Boethius later explained this as resulting from the fact that the body and soul are subject to the same laws of proportion as the cosmos.

There are some basic flaws, though, in building musical scales based solely on ratios of whole numbers. One is that using small whole ratios will cause the frequencies to be unevenly spaced in our scale. Another is that we want a scale such that if we choose an arbitrary note, we are able to play pleasing ratios of it in each octave. For example, suppose we have an A at 440 Hz. We want to play 660Hz for the pleasing 3:2 ratio, and then 990 Hz as well. But the 990Hz note might be too high depending on what instrument we are currently using, so we may want to go down an octave and play 495 Hz. We then need to add a 3:2 ratio times that 495 Hz, getting 742.5. And we will soon find ourselves in an infinite loop, having to add more and more keys to our piano and more digits to our calculations.

To solve this, musicians in recent centuries developed 'equally tempered' scales, where the frequency ratio between adjacent notes is a constant 12th root of 2. Thus, after twelve notes, you have nicely reached the next octave. While the steps in between don't precisely match whole number intervals, they do reach fairly close approximations. For example, four steps along the scale we reach a ratio equal to the cube root of 2, approximately 1.2599. This is close enough to the ratio 5/4, or 1.25, to sound pleasant when needed, even though the actual number is irrational. Rather ironic since Pythagoras's disciple Hippasus of Metapontum was executed for proving that the square root of 2 is irrational. Now that this equally-tempered scale is in common use, I'm sure Pythagoras is rolling in his grave somewhere.

Unlistenable But Fun
From Math Mutation podcast 87

Recently I've been reading a book by 20th century classical composer John Cage. He's most famous for his piece titled '4 minutes 33 seconds', which consists of four minutes and 33 seconds of silence. It was either an intellectual and artistic triumph, or a cynical attempt to demonstrate the gullibility of an avante-garde-art-loving public, depending on who you ask. I'm in the latter school, but that's not what I find most interesting about Cage. He explicitly used mathematical patterns and ideas to construct his music, and was very good about documenting his methods.

Aside from the silent piece, another famous aspect of Cage's music is the use of chance operations, or generation of ransom numbers. One method he liked to use, in compositions such as his 'Music of Changes' and 'Two Pastorales', is to decide on the notes and durations for a piece based on the I Ching, an ancient Chinese system essentially equivalent to generating random numbers from 0 to 63, though couched in a lot of mystical language. An even more bizarre method for generating randomness, used in his 'Music for Piano 1-3', created notes based on "imperfections in the paper upon which the piece was written". Even in his famous '4 minutes 33 seconds', the choice of length for the piece was determined by a chance operation.

He also liked to combine these chance operations with other types of patterns. One idea used in many compositions was to match space to time. Rather than standard musical notation to specify lengths of time, he could note at the top of each page how much space on the paper should correspond to 1 second of time. Realizing this could be confusing for musicians who tried to actually play the piece, he added notes like "Where these [instructions] are unclear, the pianist is free to decide what to do." In his composition 'Music Walk', he went even further off the deep end, giving the following instructions: "Readings are taken from a transparent rectangle having five parallel lines placed in any position over a sheet having points... The relation of points to lines is interpreted relatively with respect to any characteristic of sound or action." If that wasn't enough, in another piece titled 'Renga', he used the I Ching to choose which of a selection of drawings by Henry David Thoreau to pick out for the performer, who would then try to musically interpret that drawing, with vertical space representing pitch and horizontal space representing time. Other spatial arrangements Cage translated into music for his compositions included chessboards, with the game in progress during the actual concert, and astronomical maps of the stars.

It's fascinating to read about Cage's various methods for producing music, but a bit disappointing to hear the actual results. Actually, 'disappointing' is an understatement; in my opinion, I'd rather be listening to my cat getting vaccinated than to a typical John Cage CD. I'm not alone in this view; at one of his live performances, an acquaintance of Cage's famously got up and yelled, "John, I dearly love you, but I can't bear another minute," before walking out. I think this is the kind of music that's a lot more interesting to talk about than to actually listen to. Though I have met some serious music scholars who claim that Cage always used his own musical intuition to discipline the chance results of his operations, and that once you acquire the taste, his pieces are among the most beautiful of classical compositions. I think I'll take their word for it, and set my iPod on my David Bowie playlist while I read my next Cage book.

What a Planet Sounds Like
From Math Mutation podcast 29

You've probably heard the phrase "Music of the Spheres". You sound cool and mystical if you talk about it, but what that phrase actually means is another question. This concept, said to have come from Pythagoras and his followers in ancient Greece, is the idea that the mathematical patterns in the movements of planets and other celestial bodies form a kind of metaphorical "music". I guess mathematics and music are both abstract ways of thought that observe and experience interesting patterns. Some composers, such as Holst in "The Planets", have tried to create artistic experiences inspired by this so-called music. But the metaphor has always seemed to be a bit of a stretch to me.

Browsing the web recently, I was intrigued when I saw that a composer named Greg Fox has figured out a way to use a few mathematical tricks to literally translate the movement of the planets into music. A well-known property of musical physics is that if you look at the frequency of a musical note, the way you get to the same note in the next octave is by doubling the frequency. So, for example, the frequency of 1760 Hz is heard by us as the musical note 'A', and if we double it, to 3520 Hz, we hear another 'A', one octave higher. To the human ear, these sound like they belong together, and are in some sense the same note.

Now think about a planet orbiting the sun. Since a planetary orbit is a periodic phenomenon, it has a frequency, just like sounds in the air. Of course, it's a very low frequency. The Earth completes an orbit around the sun in 365.25 days, or 31,557,600 seconds. To get the frequency in hertz, or the number of times the Earth orbits the sun in a second, we invert this value to get about 3.169×10^{-8}. Of course, this is way below the range of sound frequencies a human ear can notice. But by repeatedly doubling it, which keeps it as the same note but in higher octaves, we eventually get about 34.025 hertz, a real musical note we can hear. In some sense, this musical note truly represents the 'tone' of the Earth's orbit.

Fox calculated the characteristic musical tones associated with each of the planets, and used these tones to construct his musical piece *Carmen of the Spheres*. This created quite a significant constraint on the music, with only 9 tones allowed, but as he writes on his website, this minimalism is the Universe's decision, not his own. Perhaps the recent discoveries of additional dwarf planets could be used to improve his music a bit in the next revision.

How does this music end up sounding? Personally, I would put it in the bucket with John Cage, Arnold Schoenberg, and Captain Beefheart, of music that's a lot more interesting to talk about than to actually listen to. But you've can download the piece online and judge for yourself. Maybe I'm just a dork who doesn't get it. But either way, it is still fun to think about the fact that the planets really do have characteristic musical notes.

Mozart Rolls the Dice
From Math Mutation podcast 54

I bet there are many of you reading this who, like me, learned to program computers as a teenager, and were fascinated when you first figured out how to make sounds come out of the speakers. If you're like me, you wrote short programs to randomly generate and play musical tones, hoping that if you ran them enough times, eventually you would produce a brilliant melody that would make you the next David Bowie. And after a while, you probably realized that randomly producing *good* music was a lot harder than it first looked. It's kind of like the monkeys-writing-Shakespeare theory: maybe it would eventually happen, but it would take a looong time.

Surprisingly, back in the classical days of the late 1700s, long before the days of computers, creating classical music through random rolls of the dice was a bit of a fad in Europe. Well-known composers including Haydn and Mozart published systems where one could roll dice to determine how to put together a large number of possible measures, and thus randomly produce a piece of classical music. They would supply a large set of pre-written musical fragments, combined with tables that showed which one to insert at which point based on rolls of a pair of dice. There are some questions about how seriously the great composers took such music though – Haydn titled his published system the "Philharmonic Joke", and Mozart published his anonymously. In fact, there is still some controversy as to whether Mozart really created his random music-creation system – the *Wikipedia* entry and another website I found seem to disagree as to whether he was truly the author.

In those days when great music was being deliberately created, why did composers feel a need to publish systems that would let people randomly come up with a musical composition? One interesting theory is that in those times and places, it became commonly expected that educated classes would be able to compose short pieces of music, just like they were expected to be able to write letters or make speeches. But needless to say, the majority of pieces created by typical people were pretty bad. By using Mozart's or Haydn's system to choose a proper set of prefabricated measures, they could at least have a shot at getting something listenable. Think of it as using clip art in a presentation instead of trying to draw everything yourself. And if the dice-roller had a little bit of talent, they could use the randomly strung-together measures as the kernel of a larger piece they completed themselves.

In the mid-20th century, these ideas of randomly created music were revived by avant-garde composers like John Cage, as I mentioned earlier. In general I have yet to hear a piece of music created by random methods that I really enjoy. Even some Cage fans I've spoken to admitted that his best supposedly random-generated pieces aren't truly random – while randomness might help inspire a starting point or escape a bout of musical writer's block, it's the inspiration and deliberate creations of the composer that create truly great music, and Cage's true talent was in manually hacking the random results when necessary. That's probably one reason why these musical dice games eventually fell out of favor in the classical period, and why even 20th century randomly generated music isn't widely discussed outside snooty art circles.

Candide's Calculus
From Math Mutation podcast 83

If you're like me, you probably recall Voltaire's satirical novel *Candide* as one of the more enjoyable 18th century novels you were forced to read in high school. Its fast-moving and rather silly plot involves a young man who is tutored by an optimistic philosopher named Pangloss. Pangloss insists that they are living in the best of all possible worlds, despite losing an eye and an ear, catching syphilis, being sold into slavery, and experiencing disasters such as a fire, earthquakes, and a tsunami. But did you know that the philosophy that Pangloss parodies is directly related to the development of calculus?

This connection comes from the fact that Gottfried Leibniz, the co-inventor of calculus, was also a well-respected philosopher. You may recall that one of the key achievements of calculus is the ability to find a maximum value of a function. This works because calculus lets us look at the slope of a curve, which measures how steeply it is rising or falling, at any infinitesimal point. When a curve has stopped rising and is about to fall, its slope is 0, and it has achieved a local maximum. So if you can calculate a point where the slope of a curve is 0, you can find a maximum.

In mathematics, this idea is not very controversial. But Leibniz extended this accomplishment into the domain of philosophy. As a basic premise, he started with his Christian religion, which asserted that there was an omniscient and omnipotent God who designed the universe. Most likely, an omniscient or all-knowing God would know calculus, and probably a much more powerful divine super-calculus than what Leibniz had developed. And being all-knowing, he would also know all the variables that would go together to describe the universe, and be able to define some infinitely complex function that would describe how good the universe is. Since God is also supposed to possess infinite goodness, it stands to reason that he would apply his super-calculus to the universe's goodness function, and achieve an overall maximum. Therefore if something local seems bad, it's only because in combination with the other variables of the universe, it needs to be that way to achieve the overall maximum goodness, the optimal result.

I find it pretty hard to argue with this reasoning, if you accept Leibniz's premises. In the centuries since Leibniz, many complicated functions have been defined, which we don't have algorithms to optimize in a reasonable time, but God would know all the mathematical techniques he needs, and wouldn't care about time limits. After all, if there is truly an all-powerful divine being who likes to create universes, he may as well take his time doing it, even if he has to spend several eons executing an impossibly time-consuming (to us!) optimization algorithm.

Thus, if your religion admits the existence of a benevolent all-powerful and all-knowing Creator, then Leibniz and Pangloss were both right, and we really do live in the best of all possible worlds.

Fractals in the Hat
From Math Mutation podcast 91

Recently I came across an interesting web article by a researcher named Akhlesh Lakhtakia at Penn State University, claiming that Dr. Seuss's classic children's novel *The Cat in the Hat Comes Back* helped him to understand fractals. If so, this would mark quite an achievement on Seuss's part, since Seuss was writing decades before fractals were named or widely studied, even in mathematical circles. But Seuss's use of repeated recursion in the book does provide some interesting analogies to fractals, so I thought it would be fun to look at Lakhtakia's argument.

To start with, let's review the concept of a fractal. At a basic level, a fractal is an infinitely complex object, usually not describable by simple Euclidean geometry, that is infinitely self-similar and generated by a recursive algorithm. The property of "self-similar" means that you can look at a small part, and it looks just like a miniature version of the whole thing. The idea of being generated by a "recursive algorithm" means that even though a fractal is infinitely complex, it is not that hard to describe. Each step in building the fractal is described by taking the previous step and making some clearly defined extension. Many phenomena in nature, such as the appearance of mountain ranges or small plants like ferns, seem to be remarkably similar to fractals, and that has led to many useful applications.

Let's make this concrete by reviewing the definition of one of the simplest fractals, the Koch snowflake, which we discussed back in Chapter 4. Draw an equilateral triangle. Now in the middle third of each side, draw another equilateral triangle jutting out, a third of the side of each original triangle. At this point, you should see a six-pointed star. Then repeat the process on each edge of this star, adding smaller triangles jutting out of each one. Repeat this an infinite number of times. Of course you can never finish drawing it, but at some point you will be drawing lines so fine they are at the limit of what you can perceive, so that's probably a good enough approximation. (Though if you're a stickler for accuracy, please wait until after finishing this book to draw your infinite number of triangles, as that might take a while.) You will see a figure that looks like a very complex snowflake.

So, how does 'The Cat in the Hat' make use of fractals? Well, in case you don't remember the story, this book centers around the Cat's attempts to clean up a mess he makes as he visits some unfortunate children. Realizing he needs some help, he takes off his hat to reveal Little Cat A, a miniature replica of himself that stands inside his hat. Little Cat A then takes his hat off to reveal an even tinier replica of himself, named Little Cat B. The hat removal continues throughout the book, until finally Little Cat Z is so small he can't even be seen. But this apparently turns out to be an asset for removing tiny hard-to-reach stains.

I think you can immediately see the similarity to the Koch snowflake: just as little triangles are recursively added to the large triangle in the fractal, little cats are recursively revealed under the Cat's hat in the book. The Cat does obey some of the key properties of fractals: the little cats are revealed through a simple recursive definition, an exact replica under the hat of the previous cat, and the self-similarity property is obeyed as well, since the process is essentially the same whether starting at the real Cat or one of the other lettered Cats. Author Philip Neil in *The Annotated Cat* points out a subtle difference though: usually in fractals, each increasingly fine level of granularity creates multiple copies of the previous level, and this is the source of the infinite complexity. In the Koch snowflake, for example, each time you are adding triangles, you are adding more triangles than in the previous step, since more and more edges have been created. Since there is only one Cat in each of the Cat in the Hat's recursive cat-creation steps, and they end at cat Z, the complexity does not really increase in the same way.

So, did Dr. Seuss really anticipate fractals? Probably not. But that doesn't mean The Cat in the Hat can't be a great metaphor for discussing the basic fractal concepts of self-similarity and recursive definitions, as well as the general mathematical idea of recursion, which comes up in both fractals and unrelated disciplines such as set theory, linguistics, and computer science. So I think Lakhtakia was right that the Cat can be a useful metaphor for introducing fractals. And of course, you don't really need an excuse to enjoy a Dr. Seuss book with your kids.

Math or Not Math?
From Math Mutation podcast 125

I recently read a surprising book by Douglas Hofstadter, the author of math-based classics including *Gödel Escher Bach* and *Metamagical Themas*. This book was called *Le Ton Beau de Marot*, and discussed the concepts and challenges of literary translation. At first, I thought it a bit odd that he would choose such a topic, which seems quite removed from his usual areas. But actually, literary translation, especially of poetry, sits at an intriguing nexus of math, language, and art.

Some of us might start with the naive view that translation is purely a mathematical process, taking the set of symbols of one language and transforming them, according to some well-defined function, into another language. But if you've gone to an auto-translation website like google.com/translate, and tried translating a paragraph from English to another language and back, you probably realized pretty quickly that there was a bit more to it. As his example, Hofstadter focuses on a short French poem called "*A Une Damoiselle Malade*", by Clement Marot, written from the point of view of a poet trying to cheer his sweetheart, who is lying sick in a hospital.

The first few lines of the poem are

> *Ma mignonne,*
>
> *Je vous donne,*
>
> *Le bonjour,*
>
> *Le sejour,*
>
> *C'est prison...*

You can see that in addition to the literal words, there is a very formal mathematical structure here: each line has three syllables, the stress falls on the final syllable, every 2 lines rhyme, etcetera. To translate it properly, you need not just equivalent words, but need to mimic this structure. In effect, you are trying to find the intersections of two functions, one that maps words to a set of corresponding words, and another that maps poetic structures into a set of corresponding poetic structures. Google Translate fails horribly, giving us the following:

> *My darling,*
>
> *I give you,*
>
> *Hello,*
>
> *Sojourn*
>
> *is prison.*

Not only are virtually all the structural elements gone, but the words just seem wrong. Perhaps the most egregious example is "*Le sejour*", translated to the out-of-place-sounding (and slightly different in meaning) English word "sojourn" instead of a more common word or phrase to match the overall tone. Not only is attention to structure needed, but a translator also needs to choose from many possible words, with different connotations and 'rings' in English.

Throughout the book, Hofstadter examines many more attempts to translate the same poem, each of which succeeds in some ways but fails in others. For example, here are a few other translations of those opening lines:

> Gentle gem,
>
> Diadem,
>
> Ciao! Bonjour!
>
> Heard that you're
>
> In the rough

That one takes a lot of liberties with the literal meaning, replacing the opening lines with gem & jewelry analogies, but provides something roughly similar and preserves the rhythmic structure. Another translation Hofstadter provides tries to put the poem in informal, contemporary language, taking yet more liberties to preserve the structure:

> Babe o' mine,
>
> Gal divine,
>
> Here's a kiss,
>
> It ain't bliss
>
> Bein' sick.

And here's another one, redirecting the poem towards Hofstadter's late wife, focusing on her final days in the hospital.

> Carol dear,
>
> Here's some cheer
>
> From your beau
>
> Lying low
>
> Has been tough

The book contains many more attempts to translate the poem, and it's fascinating to examine how in each case, different compromises are made in literal faithfulness to the words in order to make other parts of the poem more strictly compliant. We can see that the mathematical or formal process of finding corresponding words and language elements, and identifying the specific structure being targeted, is an important part of translation, but only the launching point for an artistic endeavor to truly represent the original work.

As always, Hofstadter also uses the translation of this poem as motivation for many long, interesting discussions on language, the mind, and related themes. You can see that the problems of translation would have a lot of implications for artificial intelligence: even if a computer could model internal structures that we would consider "intelligent", somehow expressing them in a way that can be appreciated by human users is fundamentally a translation problem.

He also gave me a new appreciation for both the mathematical and the artistic challenges of translation. For example, how many of us read the conductor's name with reverence whenever we attend a classical concert, but are barely aware of the translator credit when we read a book originally written in another language? Yet as we have seen, translating a book does not merely involve finding corresponding words, but requires careful artistic choice from many formally acceptable alternatives to try to be faithful to the author's intent. Arguably this is a much bigger challenge than faithfully reproducing a piece of music, which in some sense is fully specified on the written page.

Gnarly Gnovels
From Math Mutation podcast 170

If you read a lot of novels like I do, you may have noticed that that they tend to vary in quality. I know, it's shocking, but some books that are not very good manage to get published. And this was true even before the modern days of ebooks and podcasting, when any dork with a computer can publish online. (I had better not elaborate too much on that last comment.) Anyway, looking at novels in particular, what makes one more interesting than the other? Why is Kurt Vonnegut superior to a drugstore romance? It probably won't surprise you to hear that people have attempted to look at the mathematical structure of novels, and use that to analyze what makes a book good. In particular, the mathematician and cyberpunk author Rudy Rucker, in his excellent ebook of collected essays, makes several attempts to explain the quality of a work of fiction in mathematical terms.

One notion that may be useful here is the concept of Kolmogorov complexity, as applied to the outline of a book's plot. This measurement, first proposed by Russian mathematician Andrey Kolmogorov in 1963, essentially says you can measure the complexity of a string by looking at the shortest set of instructions, or computer program, that can produce the string. So, for example, a string consisting of "abababab...", repeated 100 times, can be produced by a simple instruction "Repeat × 100: print 'ab'". On the other hand, the first 200 characters of this chapter can probably not be reliably produced by any program much shorter than one that says 'print this' and lists them directly – so they are of greater complexity than the string of repeated occurrences of 'ab'.

Applied to novels, the concept is a little more abstract, but think about the plot outline of a typical schlocky romance. It probably goes something like "boy meets girl, boy loses girl, boy gets girl back", or one of a short set of variants on that concept. Thus these novels have very low complexity. On the other hand, Kurt Vonnegut's classic *Slaughterhouse-Five* is much harder to fit into a standard formula. Somehow "boy meets girl, boy gets unstuck in time, boy survives Dresden bombing, boy gets captured by aliens from Tralfamadore" doesn't seem like an adequate summary – there is still so much more to it. So at first glance, looking at the Kolmogorov complexity of the plot outline does seem to provide some insight into the quality of a work of fiction.

But in a postscript to the essay where he first proposed this measure, Rucker describes a letter from a reader that points out a major flaw in this measurement. If you accept this measure of complexity, isn't it the case that the *New York City Phone Directory* is the greatest novel in the English language? Oops! Indeed, the complexity of a phone directory, a long list of arbitrary names and numbers which essentially cannot be described much more compactly than through reading the directory itself, is astronomical by Kolmogorov's measure. And I think most of us would agree that we don't find it very fulfilling to read the phone book in our leisure time. So we need a better measurement.

To solve this problem, Rucker further refines the concept by adding a criterion for "gnarliness" of the computation required to produce a plot. Yes, I know, the surfer lingo is a bit goofy, but you have to forgive him for having lived too long in California. What he calls a "gnarly computation" is one that may have a relatively short set of rules describing it, but generates patterns of high complexity that could not be easily predicted before running the program. One example is John Conway's famous "Game of Life". (This is a mathematical game, no relation to the board game!) The Game of Life is played on an endless grid of squares, on which each square can be in a live (black) state, or dead (white) state. Each turn, any live square with exactly 2 or 3 live neighbors stays alive, otherwise dying of loneliness or crowding, and a new live square is born at any location with exactly 3 live neighbors. This simple rule, run for hundreds of rounds, can give an amazing variety of behaviors, creating long-term patterns from just a few initial live squares, seemingly living structures such as "glider guns" that endlessly reproduce, or even a universal computer. It's almost impossible to predict in advance what will happen in a few hundred rounds of the game, other than by simply running it. Thus the Game of Life, by Rucker's definition, is very gnarly.

Gnarliness, as defined by Rucker, does seem like a better measure of how interesting a novel is. It helps us rule out predictable, formulaic patterns, but also helps us rule out totally arbitrary randomness. A gnarly computation has a deep underlying pattern and structure, but you must experience the realization of this pattern to comprehend it. Even if you know the source algorithm, or the initial plot ideas that were in the writer's head, you still need to read the book to get the full effect. If the plot outline is generated by a formula so simple that once you see the outline, you don't need to read the novel, it has low gnarliness and is probably a waste of time to read. And if a novel is complex to describe but has no underlying pattern, it probably seems kind of pointless in the end.

Looking again at the example of Kurt Vonnegut's *Slaughterhouse-Five*: I think this one you can clearly put in the gnarly category. The main character has a mix of realistic and wild adventures, surviving the bombing of Dresden, getting married and attempting to start a normal life, becoming unstuck in time, and getting captured by aliens. I think we all can agree that this plotline isn't really covered by established literary formulas, and those of you who read the book realize that the quick summary I just gave you is barely a hint of the full story. Yet once you read it, the whole thing seems to fit together as a coherent whole, a classic novel exploring the insanity of war and the concept of free will. The bizarre juxtaposition of elements is shocking, but by the time you're done, there seems to be some sense to it.

Naturally, there is a lot more to judging a great novel than the math. I'm sure you can come up with a counterexample of some novel that has high gnarliness, but still stinks. And there is some level of looseness in this definition of gnarliness, as the plot of a novel isn't really a precise mathematical object in the same sense as a mathematical construct like the Game of Life. There can also be honest disagreement on the level of gnarliness of a novel: ironically, having read many of Rucker's cyberpunk novels, I think he himself occasionally departs from true gnarliness to near randomness. On the other hand, if you ever do decide to create a work of your own, this is a nice guideline to keep in mind: if you want to hold readers' interest, it's a very good idea to strive to keep an underlying pattern discernible to the reader, but make it as gnarly as possible.

CHAPTER 9

■ ■ ■

Political Ponderings

The decisions about how human society is to be organized and governed are among the most critically important choices a civilization must make. Thus it is only natural that we should attempt to use insights from mathematics, which enables precise reasoning based on well-established premises, in order to improve our decisions in this area. On the other hand, this is easier said than done, as any political decision is heavily influenced by value judgements that are beyond the domain of mathematics – despite the fact that one side or another will dogmatically cling to them regardless of evidence or of outside events.

I have generally avoided any topics in *Math Mutation* that would imply a strong political stance on a controversial issue, in order to avoid alienating a subset of listeners in today's polarized climate. But I have personally ventured into politics at the local level, successfully running for a seat on the board of Oregon's fourth largest school district, so political thoughts are never too far from my mind. Thus, on numerous occasions, *Math Mutation* has included topics that touch on politics in one way or another. I hope you will enjoy the historical anecdotes, and perhaps you will agree with some of the ideas you read in this chapter. I still believe my proposal for solving the problem of endless recounts, which you will read in one of the sections below, would be a major boon to the American election system!

A Founding Theorem
From Math Mutation podcast 183

If you're a fellow American, you probably clean off your barbecue and put out your flag every July 4th for the holiday. This is the day set aside once per year to celebrate the independence of the United States from England, and to reflect upon the mathematical theorems that led to the founding of our nation. What? Are you thinking that July 4th isn't about math? Clearly something has been missing from all the specials you've been watching on TV. Actually, there was at least one mathematical result that had a direct influence on our Founding Fathers: Condorcet's Jury Theorem. This theorem states that if you are trying to decide on a topic by voting, and the average voter has at least a 50-50 chance of getting it right, increasing the number of voters gives a more accurate result.

Back around the time of America's founding in the late 1700s, there was a colorful French mathematician, the Marquis de Condorcet, who spent a lot of time thinking about mathematical aspects of democracy, voting, and probability theory. Condorcet is known to have collaborated with colleagues in the US and worked personally with Benjamin Franklin, and thus his theorem is thought by many to have directly influenced the United States Constitution of 1787. The Jury Theorem was part of a larger 1785 work, *Essay on the Application of Analysis to the Probability of Majority Decisions*, considered a major contribution in the development of probability theory. Condorcet was concerned both with the basic concepts of probability, and how they could be used to help develop rational public policies – he was one of the earliest mathematicians to attempt such a direct application to the social sciences.

To get an idea of how we prove that increasing the number of voters leads to a more accurate total, think about a situation where we have some odd number n of votes, the probability of an individual vote being correct is p, and we add two more votes that might change the result. In this case, you can think of the nth vote as having been a deciding vote: our oddness assumption prevents ties, and otherwise the two new votes would not have been able to reverse the answer. So we just need to calculate probabilities for the last 3 votes, in 2 cases: did they convert an incorrect majority to a correct majority, or a correct majority to an incorrect majority? Remember, each individual vote has a probability p of being correct and $1-p$ of being wrong. Also remember that you can compute the overall probability of a set of independent events by multiplying their probabilities together. So for the previous deciding vote to be incorrect, and the two new votes reversed the tide, the probability would be $(1-p) \times p^2$. For the previous vote to have been right, and the two new ones to reverse the total, it would be $p \times (1-p)^2$. The first one of these is larger than the second if and only if p is greater than 1/2.

We should be careful, though, to look at the limitations of this theorem before relying on it too much in our political system. At its heart is the notion of some kind of objective correctness, the ability to state a right answer that most people will come to with a definite probability. Except when I'm one of the candidates, there are very rarely such clearly correct answers in politics. It also depends on the voters being independent, and ignores the influence of neighbors, popular delusions, and people too lazy to gather the information needed to make an intelligent decision, or people held in sway to demagogues. I'm pretty sure that, depending on your political leanings, you consider the election of one or more of our past two U.S. presidents a result of such factors. And the theorem only works when choosing between precisely two alternatives: it does not account for voting on multiple choices.

The last limitation leads to another of Condorcet's important insights, his "voting paradox". Assume we have 3 opponents A, B, and C, and the public is divided into 3 segments with cyclic preferences: some prefer A over B over C, some want B over C over A, and some want C over A over B. Let's say they vote and get basically a 3-way tie, with just a very tiny margin putting C over the 1/3 number needed to win. You could argue that if C won, then 2/3 of the population would have wanted B to win over C: since the ones that wanted A preferred B over C, and the ones that wanted B also preferred B over C. Yet if B won, you could say precisely the same thing about the people preferring A, or if A won, could make a similar complaint about C! Thus, somehow a popular vote will always result in an outcome that 2/3 disagree with. This can be solved by voting with a "Condorcet method", basically a more complex voting method that guarantees the winner will be someone who would have won in a pairwise competition with every other candidate. This can be done by having voters list candidates in order of preference instead of just choosing one result, or by having multiple rounds of voting with runoffs. Personally, I think these kinds of alternative methods make a lot of sense, though they suffer from the disadvantage of being much more complex to administer and understand.

So, with all these great insights into how government should work, you would infer that Condorcet had a brilliant political career, right? Unfortunately, he suffered the results known since Socrates to those who embrace their philosophical insights too enthusiastically without sufficiently understanding the flawed human psychology all around them. Initially a supporter and active participant in the 1789 French Revolution, he failed to support the radical faction that took control, instead trying to convince them to embrace his ideas. While this got him a position on the Assembly for a while and a decent amount of popular support, it eventually led to his arrest and death. Technically he killed himself in jail, but many believe this was permitted or staged to avoid the embarrassment of publicly executing such a well-known revolutionary figure. Either way, in the most important applied mathematics experiment of his life, the basic assumptions of Condorcet's theorems just didn't hold. Let's hope my term on the Hillsboro School Board leads to a better fate.

More Than a Cartoon Cat
From Math Mutation podcast 88

If you paid attention in high school history class, you may vaguely recall the name of James Garfield, the 20th president of the United States. In his early life he had been more academically inclined, but the Civil War interrupted his plans. He turned out to be a talented soldier, and quickly rose through the military ranks. Afterwards, as a popular Civil War veteran, he was elected to Congress and then to the Presidency in 1880, the only president in history to have been elected while sitting in the House of Representatives. Tragically, he was assassinated within four months of being sworn in, so didn't have much time to get himself noticed in our history texts. But he is remembered for one very unusual achievement: he is the only U.S. president known to have published a journal article in mathematics. He's credited with an original proof of the Pythagorean Theorem, created while he was sitting in the U.S. House of Representatives.

As I'm sure you recall, the Pythagorean Theorem is the classical result that if a right triangle has legs of length A and B and a hypotenuse of length C, $A^2 + B^2 = C^2$. Garfield's proof worked by constructing a particular trapezoid. Place two copies of your right triangle next to each other in a straight line, touching at the tip, such that the length-A leg of one is next to the length-B leg of the other. Now connect the upper vertices of the two triangles. You should now have a trapezoid whose base is of length $A + B$, and has height of B on one side and A on the other. If you draw this figure, you will see that the trapezoid consists of three triangles: the two you started with, plus one in the middle that was formed when you connected the tops – and that new one is a right triangle whose legs are both of length C.

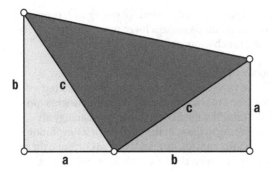

Figure 9-1. *Garfield's trapezoid*

Now let's compute the areas of the three triangles. The two original triangles are right triangles with legs of size *A* and *B*, so their areas are *AB/2*. The middle triangle has area $C^2/2$. But now let's equivalently compute the area of our full trapezoid, which is the base times the average of its sides, or $(A+B) \times (A+B)/2$. Then we just need to equate our two computed areas: $AB/2 + AB/2 + C^2/2 = (A+B)^2/2$. Simplifying, we get $A^2 + 2AB + B^2 = 2AB + C^2$. If we eliminate the *2AB* terms, we have our theorem.

I should point out that this isn't really an earth-shattering mathematical result. One of the classical proofs involves constructing a giant square using four copies of the original triangle, and Garfield's proof essentially constructs half of that giant square and uses trapezoid-based instead of square-based areas for its calculations. But I still think it's nice that at least one U.S. president did make a real contribution to mathematics. Given what's been going on with the federal budget, there may be doubts that any of the recent ones can do basic math.

The Round Road to Damnation
From Math Mutation podcast 20

Every schoolchild, and probably most listeners of this podcast, knows about the constant called 'π'. This is the ratio between the circumference and diameter of a circle, an infinite decimal beginning with 3.14159. Those of you who are somewhat religious Judeo-Christians may thus be concerned when you hear that there is a passage in the Bible's Book of Kings where it says a round basin has a 10-cubit diameter and a 30-cubit circumference, implying a value for π of exactly 3. Does this mean you have to choose between your religion and your math?

Fortunately, there are several ways to reconcile this Biblical story with reality. The simplest is to say this is just an approximation; after all, the Bible was written many years before the decimal system, making a potential detailed description of the measurements rather cumbersome. A famous rabbi named Nehemiah came up with a more clever explanation around the year AD 150: the diameter was measured using the outer rim, while the circumference was measured using the inner rim. This way, depending on the thickness of the basin, the values could very well be made correct to any desired degree of precision.

Another interesting element of this confusion is the persistent stories going around the net that crazy Christian fundamentalists in the U.S. Midwest have tried to pass laws making π equal to 3. These stories have been around long before the internet, actually – Robert Heinlein mentioned this in passing in one of his novels. This always sounded a bit suspicious to me; after all, not only was π known to within 0.5 % accuracy by the ancient Babylonians, but you can also disprove the π = 3 theory with a dinner plate and some string, even if you're an uneducated farmer in the remote countryside.

The consensus of urban legend-type sites seems to be that this rumor is false, and Christians never tried to pass such a law. The bit of truth that started these stories is a proposal from 1897 in the Indiana legislature to change the value of π. The proposal, originally proposed by a rather deluded pseudo-mathematician named E.J. Goodwin, claims that 3.14159 is "wholly wanting and misleading in practical applications." Apparently the bill was so incoherently written that it wasn't even self-consistent, proposing values for π of 3.2, 3.23, or 4 at various times. The bill actually passed the state House, but was held up indefinitely in the state Senate for unclear reasons, and never became law.

So, what's the lesson here? Well, people in Indiana may be crazy, but not because they are Christians. And you can continue to correctly measure the area and circumference of circles without fears that you will suffer for it in the afterlife.

A Math Teacher to Remember
From Math Mutation podcast 93

With my mind blanking on podcast topics one day, I looked through one of those online lists of people who died recently, to see if there was anyone of mathematical significance to mention in the podcast. One name that stood out was Aleksandr Solzhenitsyn, the Soviet dissident who was instrumental in publicizing the horrors of the communist system to the outside world. What is lesser known about him is that he was originally trained in mathematics, and this background played a key role in enabling his later political activities.

Solzhenitsyn was born in 1918, so came of age just in time for World War II. While having an interest in literature from an early age, he found in school that he had a talent for math, so studied at the Department of Mathematics in Rostov University, graduating in 1941. Because of this background, he was sent to an artillery school, and served in the war as an artillery officer until he was arrested for having made remarks critical of Stalin in a private letter.

Upon his arrest, Solzhenitsyn was deported into the Soviet Union's "Gulag", or network of concentration camps, where political prisoners were generally put on heavy labor duty with barely any food or shelter, until many were worked to death. But his mathematical background saved him: he was one of the few lucky prisoners selected to serve in the *sharashkas*, a kind of university system within the prison camp network. Those prisoners were given a better level of food and shelter, and expected to conduct advanced scientific research.

In his autobiographical novel "The First Circle", Solzhenitsyn talked about life in these camps. He related some darkly humorous anecdotes about the prisoners constantly trying to convince their technically illiterate Communist masters about progress being made. In one example, they convinced senior officials that they could look at a voiceprint and figure out what was being said, by of course colluding in advance about what statements to use in the test. But the stakes were deadly serious, as any prisoner in whom the authorities lost confidence would likely end up back in the common labor camps.

Eventually Solzhenitsyn was declared 'rehabilitated' and released, and once again his mathematical background played a key role. A "rehabilitated" political prisoner was still living under many restrictions and had very few choices in where or how to live. But because he qualified to serve as a math teacher, he had had a relatively comfortable existence for an ex-prisoner, with enough time to write. And then he skillfully read the political winds and submitted his novel about prison camp life, *One Day in the Life of Ivan Denisovich*, at just the right time: Krushchev, the new Soviet leader, was trying to show his independence from Stalin's legacy, and allowed it to be published, despite its stark exposure of the inhumane treatment of prisoners. The Soviet leadership came to regret this decision, especially after Solzhenitsyn won the Nobel Prize and became too famous to quietly dispose of like so many others who had criticized the communist system.

Since this is a math podcast rather than a political one, I'll avoid talking in too much detail about Solzhenitsyn's many important political works, which exposed not only the treatment of political prisoners, but the many ways in which the communist system destroyed the moral, political, and cultural fabric of Russian society. If you have any interest in the topic, I do highly recommend that you pick up *The Gulag Archipelago*, his massive work based on the collected recollections of himself and the thousands of other prisoners who wrote to him after he became famous. You would think 1800 pages on life in Soviet prison camps would be dry, but once I started reading it, I couldn't put it down.

Election Solutions
From Math Mutation podcast 172

Close elections, where two candidates have vote totals that are close enough to be within the possible counting error, seem to be a perennial feature of our voting systems. Nearly every U.S. election these days is accompanied by stories of various groups of lawyers gearing up to challenge totals and force recounts in various states or counties. You can probably remember many stories about nail-biters like the 2000 US Bush/Gore presidential election, the 2008 Minnesota Coleman/Franken senatorial election, or the earth-shattering tension created by the 2009 Hillsboro, Oregon Seligman/Sollman school board vote. But do endless hand-wringing, legal battling, and recounts after close elections really make logical sense? Do they improve the quality of our election results? Recently I've been reading Charles Seife's entertaining but occasionally flawed 2010 book, *Proofiness: How You're Being Fooled by the Numbers*. Seife presents an interesting analysis of these situations from a mathematical point of view.

Seife describes the core issue here as what he calls "disestimation". What is disestimation? Here's a particularly egregious example: A museum tour guide was showing some visitors the skeleton of a large dinosaur. One of the visitors asked how old the skeleton was. The tour guide replied, "Exactly 65 million and 38 years old." When asked why he was so confident, the guide replied, "I started working here 38 years ago, and the paleontologist told me this skeleton was 65 million years old at that time." As you can probably see, the guide was ludicrously misunderstanding the precision with which scientists can age fossils: what was intended as an estimate within a few percent of the real total, he took as a precise statement of an exact value. "Disestimation" is the tendency to incorrectly treat an estimated value as much more exact than it truly is.

Similarly, whenever we get the total results from an election, it is an estimate, NOT an exact count. So rather than saying that the 2000 US Presidential vote totals in Florida were "Bush 2,912,790 / Gore 2,912,253", it would have been more correct to say that they each got approximately 2,910,000 votes The nominal totals are simply an estimate, just like the dinosaur's age. At first this sounds a bit odd – aren't votes an exact numerical total that can be confidently counted? In fact, when counting large numbers of objects, it's pretty well established that humans will make occasional errors. It's conventional wisdom among factory managers that if you ask two employees to take inventory by counting the objects in a warehouse, you'll get two different results. In the show notes you'll see a link specifically to a Rice study on hand-counting votes, demonstrating an expected 1–2 % error. I don't think many will disagree that even when computers are used in some degree, you have issues like paper jams, power flickers, mistyped human input, or software bugs. With these kind of error rates, a recount is really just a flip of a coin: the new result will be different, but just as error-prone as the original result.

So, what should we do when we have a close election? Seife proposes a simple answer: rather than going to the time and expense of a recount to achieve an essentially random result, let's agree that when the election is within a certain margin, we will declare it a tie, and flip a coin to decide the winner. The result will be no less accurate than the one achieved by the recounts, and will be achieved with much less time and expense. It's a simple, elegant solution – but Seife misses one fatal flaw in this proposal. What is the exact margin that is required to declare the election a tie? And if we're close to that margin, how do we decide if it's really a tie or not? For example, in the Bush/Gore Florida vote, suppose we had calculated that 50.005 % of the vote would constitute an outright victory, which would have put Bush's total 12 votes short of the amount needed to avoid the coin flip. This would have resulted in the same drawn-out battle, forcing recounts to see if the total could be pushed to the result that would force or avoid the coin flip. By the way, this type of situation is not purely hypothetical – a few years ago the race for the Oregon Superintendent of Education was in a very similar situation, where Ron Maurer lost by an amount just slightly greater than the margin that would trigger a statewide recount. Supporters tried to push for local county recounts in close areas, hoping for a result that would just barely push him into the boundary that would trigger the statewide recount, essentially a gigantic coin flip. This was despite the fact that there was no way that recounting those votes could give Maurer an actual majority.

I think the key issue Seife missed is that we will have this problem any time there is a discrete boundary where one vote, or a small handful, makes a huge difference in the outcome. Is there a solution to this? I see one possible method. We could use a meta-solution based on Seife's initial solution: if the margin is close to the value of what would trigger the coin flip, by a certain percentage, we have two coin flips: one to figure out if we should consider the original vote to be within coin flip range, and then a second to actually decide the election if the first one succeeded. This would effectively give the candidate with the lower total a 1 in 4 chance of flipping the election. But of course, then we have the meta-meta-issue, of what to do if the vote total was just outside the margin that would cause us to flip a coin to see if the total was within the margin that we flip a coin for the election. Now we would have 3 coin flips, with the lower original vote-getter having a 1 in 8 chance of flipping the election. And what if we were just outside the margin of the margin of the margin? That can be solved by 4 coin flips... And so on.

One other wrinkle here though: since we are calculating percentages of percentages of percentages, the total of the margins we are looking at is effectively a geometric series: $1/n + 1/n^2 + 1/n^3...$, which means that the total of all our margins will converge to a known value, beyond which we need to do no coin flipping. For example, if we allow a 10% margin to trigger the first coin flip, a margin of $1/10 + 1/100 + 1/1000...$, which equals 1/9 of the total vote, will be the outer boundary of where a series of coin flips would occur. Will the discrete 1-vote boundary between coin-flipping and non-coin-flipping itself recreate the original problem? It's exacerbated by the fact that you can't have fractions of a vote, meaning that the series of margins would peter out after 7 or 8 iterations, when the size becomes less than one vote. The losing candidate would probably go to great lengths to cross this boundary if close & give himself a remote but possible shot, again recreating all the legal wrangling. Perhaps it would be better to have some continuous sliding scale of probability based on a bell curve: if one side got 99% of the vote, we have a 1-in-a-billion chance of reversing the election, and we gradually scale this down based on the vote %, smoothly reaching the coin flip at 50% of the vote without having any specific point where 1 vote causes a huge, discrete jump in the outcome. For any particular vote count, there would be a known probability of reversing the vote, and slight increases or decreases to the vote count would just result in small changes to that probability.

With this system, we would have a huge return in societal resources that would not be wasted on endless recounts. With all the financial and budget crises today, think about what this means: how many teachers' jobs could be saved if the salaries of all of Obama and Romney's recount lawyers were instead donated to schools? I think I'll call up President Obama and ask him to start working on it. Unfortunately, this system would not survive the first lucky win: when some popular candidate gets a clearly higher vote total but loses due to a lucky one-in-a-thousand set of coin flips, which is bound to happen somewhere given the number of state and local elections nationwide, there will probably be riots in the streets and popular pressure to go back to the system we have now. Oh well, I guess that's why I'm a podcaster and not a president.

Democracy Doesn't Work
From Math Mutation podcast 188

After the work by Condorcet mentioned a few sections back, work on mathematical analysis of election systems continued until the preset day. One of the most famous results of the past century in this area is Arrow's Theorem. Arrow's Theorem was first proven by economist Kenneth Arrow in 1951, as part of his Ph.D. work – just the start of a long career that later won him a Nobel Prize. This is a theorem that basically says that according to some common criteria that we should use to define a fair voting system, no rank-order voting system can ever meet those criteria. In some ways it can be considered an extension of Condorcet's voting paradox. I think the easiest way to introduce Arrow's Theorem is through an anecdote I read in the *Cafe Hayek* blog by Don Boudreaux.

"You walk into an ice-cream store and ask what flavors are available today. The clerk says 'We've got vanilla and strawberry.' You ponder for a moment and tell the clerk 'I'll have strawberry.' Just before the clerk starts to scoop out your strawberry ice cream, he turns to you and says, 'Oh, I almost forgot. We also have pistachio.' In response, you ponder for another second and then tell the clerk, 'Well, in that case, I'll have vanilla.'

Seems pretty absurd, right? Somehow the availability of a flavor you don't like changes your first-choice selection? But when we're talking about voting, our society really does choose like this. In one recent election, people pointed to the Virginia governor's race, where many believe that Republican Cuccinelli only lost to Democrat McAuliffe because of the presence of 'spoiler' Libertarian candidate Robert Sarvis. McAuliffe won the election, but if Sarvis had not been running and his voters had chosen Cuccinelli (whose views were very close to Sarvis on most major issues), then Cuccinelli would have won. Even if you disagree in this instance, articles on the spoiler effect appear during pretty much every election season in the U.S.

This spoiler effect is related to one of the criteria for fair voting in Arrow's Theorem, the "Independence of Irrelevant Alternatives" – the idea that if you prefer X over Y, your feelings about some third alternative Z should not change that. There are three other criteria in the theorem. First, "non-dictatorship": no single voter should be able to decide the outcome. Second, "universality": for any set of votes, the system must provide a complete, deterministic ranking of the society's preferences as a whole. Third, "unanimity": if every individual prefers one choice over another, then so must the society as a whole. The theorem then states that if you have at least two voters and at least three options to decide among, it will never be able to meet all of these four fairness criteria.

As you would expect, Arrow's theorem has led to lots of discussions about how to improve democratic voting. One simple way is to cheat and just relax one of the fairness criteria – in fact, this has largely been done in practice, as we do have voting systems in many countries, including the United States, that do allow the spoiler effect and violate the independence of irrelevant alternatives. Another method is to always limit votes to two alternatives, since then we can have a 'fair' system according to the Arrow criteria – but unfortunately if we divide a larger group of alternatives into pairs to try to use this method, then we find the collective choice using these multiple pairwise votes is in effect a larger tournament that is subject to Arrow-like problems. The order of pairing can have a big influence on the ultimate winner. There are also systems not based on rank order, for example 'Range Voting', giving a score to each candidate instead of a simple rank order and adding the populations' total ratings: this has its own problems though, sometimes giving a result close to society's average judgement but disagreeing with the true majority choice.

Some mathematicians have also pointed out that if you drop the assumption of finitely many voters, Arrow's theorem can be fixed, but I think there might be a few other problems if we increased our birth rate to infinity just to fix our voting system. Plus, until they reached voting age, we wouldn't be able to fairly elect a school board to oversee the education of our infinite number of children anyway, so they would not grow up to be informed voters.

All these improved voting systems, or at least the set of them that are actually possible in real life, suffer the disadvantage that they make voting more complicated in general. Given the contentiousness and error rates we have now regarding the simple problem of counting direct votes, I think that is likely a fatal flaw. At some level, we just have to recognize that any system of governance will have its inherent flaws, and that we just need to learn to cope with the fact that life is never 100 % fair. And the fact that after every election, advocates for the losing side will come out and declare society fatally flawed, is just one of the prices we pay for the public having some kind of influence on its government – every fix they advocate will result in some other form of unfairness.

Personally, I still say the best system would be the one I proposed in the last section, where we simply admit at some point that the election system is uncertain, and roll a die to help randomly determine the final result, with probabilities determined by the votes counted so far. Or at least, that would be the best stand-in until society gains the wisdom to make me its benevolent philosopher-king.

Drop That Number or We'll Shoot
From Math Mutation podcast 60

Would it surprise you to learn that there are certain situations in which writing down a mathematical theorem, or even a particular number, can be a violation of the law? This odd situation results from the interactions between copyrights, trade secrets, and American law in the modern world. Under the U.S. Digital Millennium Copyright Act, or DMCA, originally passed in 1998, virtually any attempt to circumvent protection of copyrighted works, including figuring out any encryption methods, is against the law. Many researchers were unhappy about this, as research into encryption is a vital and active area of modern mathematics, and under some interpretations, much of it could be considered illegal. Probably the most famous specific controversy that came up related to this law was the publication of the DeCSS encryption algorithm, which was used to encode video DVD files.

David Touretsky, professor of computer science at Carnegie Mellon University, made a strong case that the DeCSS algorithm belongs to the realm of mathematics and of free speech, and should not in itself be considered an illegal device. To support his case, he came up with a really silly gallery of different ways of expressing the DeCSS algorithm. Here are a few:

- The most basic example is the original computer program that directly decodes DVDs, obviously the most legally shaky case.

- Next there is a computer program written in what is known as a "functional language". A functional language is essentially a computer program that doubles as the statement of a mathematical theorem. Can it really still be illegal?

- Then, taking it a step further, the gallery contains a pure mathematical theorem proving that DVDs can be decoded. The decryption algorithm is described within the proof, without providing actual computer code to do so. Can it really still be illegal?

- Moving off in an artistic direction, the gallery also contains a haiku poem which states the essential properties of the algorithm, in carefully structured syllables as prescribed by ancient Japanese artistic techniques. Along similar lines, there is a dramatic audio reading of the algorithm, including musical and square-dance variants. Can you be arrested for dancing to the tune of an illegal theorem?

- Then there is an artificially constructed DNA sequence that encodes the algorithm. With recent technological developments, maybe we could actually build an illegal bacterium.

- Another example is the board from a game of "minesweeper", in which the patterns of mines describe the DeCSS algorithm. I guess as long as you get blown up before uncovering all the mines, you won't violate the law.

- An encoding of the algorithm as a single prime number. Could this be the first case in history of a particular number being illegal? Can you go to jail for numerical possession?

You can see the full gallery at Touretsky's site online. It's pretty fun to come up with lists like this, but I do think Touretsky was glossing over a serious issue. I'm not a lawyer, but if you take his arguments to an extreme, it seems to me like you have to throw away all concepts of copyright, trademark, and patent: with modern digital encoding methods, everything stored digitally anywhere is a string of 1 s and 0 s, so can literally be viewed as just a very large number. I think we do owe a lot of our modern technology to the fact that people can come up with ideas and expect they will get some kind of legal protection; are we really ready to toss all that out in order to regain full mathematical consistency?

In the end, Touretsky didn't convince the judge, and the DMCA remained in effect. So maybe we do have to accept a world in which every once in a while, math really can be illegal. But don't worry–if I end up in jail, I promise to find a way to keep this podcast going.

That's How We Do It In Government
From Math Mutation podcast 158

At a meeting of the Hillsboro, Oregon School District curriculum committee, discussing progress in English Language Learner education, a district official presented a statistic that looked something like this:

Mean test score: 73 %

Margin of error: 9 %

Adjusted score: 82 %

Now, for any of us in engineering or other professions that use margins of error, this looked distinctly odd. A "margin of error" represents the imprecision in a measurement, and inherently can show uncertainty in either direction, at some specified level of confidence. So a rating 73 % with a 9 % margin of error might mean we are 95 % confident that the true score has a range of 64 %–82 %. It's equally likely that the valid result is at either end of the range. Why does it make sense to add the margin of error to the mean score, calculating the maximum score at the high end of the confidence interval, when reporting the "adjusted" official result? The answer, when I raised the question: "That's how we do it in government."

This is a nice trick: it enables every statistic to be presented in the best possible light for promoting the success of current public officials. It also is inherently insane, in my opinion, granting bonus points for the imprecision of the measurement. Think about it: normally, measurements with lower margins of error are seen as more valuable, as they give a clearer and more precise picture. But look at the scores above: if they worked hard on developing a better test and lowered their margin of error, the "adjusted score" would likely be penalized! And if they know the true scores are going down, they can game the system by lowering the quality of the tests or sampling, aiming to increase the margin of error rather than improving student knowledge. Is this the right way measurements should be done in our education system?

I can see how this would become the custom in government: once one official does it, everyone else has to follow suit, or else their statistics would appear inferior. Imagine if the district suddenly stopped "adjusting" these scores. "Look, in Hillsboro the scores went down 9 % this year!" Any elected officials involved would see their opponents demagogue the issue, and the employees who stopped the adjustments would suffer for it.

Don't take this post as a criticism of the particular official who made this presentation though: in fact, I am commending him for his forthrightness when explaining the topic to me. In a regime where this silly "adjusted score" must be produced, the most intellectually honest policy is to do what he did: present the actual source numbers in addition to the final adjusted score, and let the viewers see the full story. I'm happy to see our district doing this.

The big lesson: any time a government body reports an "adjusted" statistic, look very closely at the adjustment, and demand to see the raw scores if not shown.

CHAPTER 10

■ ■ ■

Money Matters

Money and economics are among society's earliest motivations for learning about and developing mathematics. If you think about it, the fact that we are able to facilitate intelligent business and exchange policies is not something we can take for granted. Isn't it somewhat miraculous that one person can spend 40 hours a week designing computers with no awareness of where his food comes from, another can spend the same amount of time farming with no awareness of what goes on at the computer factory, and yet both end up with easy access to food, computers, and unrelated services like medical care? Somehow the glue of economics, aided by its mathematical foundations, enables this to smoothly happen in our modern society. In this chapter we highlight some fun historical anecdotes and some more modern thoughts on concepts related to money, economics, and business, intertwined with various aspects of mathematics.

Tally Folly
From Math Mutation podcast 96

These days, we are so used to universal literacy and ubiquitous writing materials that many of us never even think about how we would count and keep records without writing. In fact, until a couple of centuries ago this was a serious problem, and many people and businesses kept financial records in the form of tally sticks, sticks with notches marked to represent various quantities. Use of such sticks extends far into antiquity: the oldest known mathematical artifact is said to be the "Lebombo Bone", a 35,000-year-old baboon fibula with 29 distinct notches found in Swaziland that resembles calendar sticks still used by bushmen today. Even in Western countries, the use of this method lasted longer than you might have thought. For example, the move from tally sticks to paper records in England didn't occur until the early 19th century. Some stories even claim that early New Yorkers originally located their stock market, now Wall Street, to be near a plentiful supply of wood in case it was needed for tally sticks.

The system of keeping records with tally sticks used by the Exchequer, or tax authority, of England had been handed down since the 12th century, and was actually pretty clever. First notches were made of various thicknesses to indicate a numerical quantity: a thousand pounds was a cut as thick as "the palm of the hand", 20 pounds the "breadth of the little finger", etc. Then the stick would be split in half lengthwise, and each party to the transaction given half the stick. Because it would be nearly impossible with premodern technology to construct a fake replica of your half of the stick with different notches, this was a nearly tamper-proof method for recording taxes owed or similar transactions. Tally stick halves were accepted as legal proofs in European courts and under the Napoleonic code. And a tally stick half held by the Exchequer, which showed an amount of tax money owed to the king, could be sold by the government when cash was needed before taxes were due. Thus notched tally stick halves actually functioned as an early version of a short-term bond, beginning a tradition of innovation in the British monetary system.

Of course, times moved on, and English treasury officials had been trying to move off the tally system since 1724. But due to popular usage and momentum the sticks had to be accepted until 1826. By 1834 the rotting stock of old tally sticks stored by the government at Westminster was getting to be rather annoying, so they tried to work out a way to free the space. Charles Dickens pointed out that they could easily have disposed of these sticks and put them to productive use by simply allowing London's poor to take them for use as firewood: but as usual when a government was involved, the official policies dictated a different procedure, and the old tallies had to be burned in an official stove in the House of Lords.

If that wasn't dumb enough, there was the further problem that there were way too many tally sticks to fit in this stove. And the workers in charge of the burning were not paid to think, but to obey, and according to some stories were concerned about finishing the job by the 5pm deadline. They stuffed in as many tally sticks as they could, and lit the flame. Dickens describes what happened next: "The stove, over-gorged with these preposterous sticks, set fire to the paneling; the paneling set fire to the House of Commons; the two houses [of government] were reduced to ashes; architects were called in to build others; and we are now in the second million of the cost thereof." In other words, Parliament was burned to the ground, and had to be rebuilt from scratch.

How to Bankrupt Your Boss and Get Rich
From Math Mutation podcast 84

With all the economic turmoil in the news lately, I've been leafing again through my copy of *Fooled by Randomness*, Nassim Nicholas Taleb's excellent book about basic probability fallacies that impact modern investors. One important topic introduced in that book is the concept of asymmetric bets. It's not a very complex concept mathematically, but it's still one that a surprising number of people fail to think through when investing.

Suppose I were to tell you that mathmutation.com stock has a 90% chance of going up in the next year, and only a 10% chance of going down. For the moment assume we have completely accurate information, so you are sure that these probabilities are correct. Your instinct might be to go out and buy a bunch of shares. After all, there is a 90% chance that you are going to make a profit. But is this the correct answer? Actually, you still need more information to figure out whether the stock is something you should buy or should sell. To see this, let's assume the stock value is currently 100 dollars, and look at the expected profit from buying a share in two scenarios.

In scenario one, if the stock either rises or falls, it will be by fifty cents. So if we buy a share, our expected gain is $.9 \times 50 + .1 \times (-50)$ cents, for a total of $0.40. Thus, due to the positive expected gain, this is a buy – in the long term, you expect on average to gain money when buying in this scenario.

Now let's look at another scenario. Here, the 90% chance of a rise is due to slowly spreading word-of-mouth about *Math Mutation*, so if the stock goes up, it will still be about fifty cents. However, the 10% chance of a fall is due to rumors that the host will soon convert to the Amish religion and give up all technology, so there will be no further episodes and the podcast will be worthless, and the stock will drop by the full 100 dollars, or 10000 cents. Then your expected gain will be $.9 \times 50 + .1 \times (-10000)$, for a total of -\$9.55. In the long term, using this strategy repeatedly and deciding to buy the stock, you expect a huge net loss.

This is an example of an asymmetric bet – while there is a larger chance of a gain, the magnitude of the gain is not enough to justify the risk of the huge loss. Even though there is a 90% chance of the stock going up, the one in ten times it goes down will be such a catastrophe that it will wipe out all your previous gains, and more. So it seems like an intelligent investor should always try to recognize such asymmetric bets, and make the rational decision not to invest if there is a large chance of a small gain, but a small chance of an overwhelming loss like this, right? Well, we need to add one more twist.

Suppose you are managing a huge corporate portfolio worth hundreds of millions of dollars. Each year in which the corporate account gains, you get a half-million dollar bonus, which you can then put in a personal savings account. However, if the corporate account goes bankrupt, you will be fired. Now think again about the asymmetric bet. If there is a 90% chance that it will gain, then you know you can place this bet for your company – nine out of ten years, you will get a huge bonus. Eventually you will probably bankrupt the company and get fired, but by then you will have squirreled away enough bonuses from your successful years that you won't care. In fact, you probably have a decent-looking resume for applying to further financial jobs, with a record superficially showing that most of the time, you made money for your company.

I don't know enough details to say whether this is what actually happened in any of the recent crises, and I'm sure economists siding with one company or government party will disagree with those siding with rivals. In his book, Taleb does describe many examples of individual traders who seem to be doing well for a number of years, then suddenly lose their companies a huge amount of money that dwarfs the total of their previous gains. And recent additions to Taleb's website seem to indicate that he thinks his ideas applied directly to Fannie Mae, a semi-private U.S. company that many blame for the recent economic turmoil. In any case, in the current situation, it does seem suspicious to me that there are so many cases where individual executives exited with millions of dollars, while leaving behind bankrupt companies for the taxpayers to bail out.

What Color Is Your Swan?
From Math Mutation podcast 164

In the past week, a popular topic of conversation here in Hillsboro has been the quarterly earnings announcement of our largest local employer. As with all companies, the quarterly earnings affect company stock, bonuses, and lots of other related arcane financial statistics. But are short-term earnings a good way to judge a company? Could this kind of result be inherently inaccurate or even deceptive? I've recently read Nassim Nicholas Taleb's book *The Black Swan*, which offers a unique perspective on these kinds of measurements, with much wider implications throughout human society.

In Taleb's terminology, a "Black Swan" is an event with low but nonzero probability, hard or impossible to predict based on current knowledge, whose effects are large enough to dwarf many previous high-probability events. A good example in the corporate world, which we also discussed in the previous section, is the case of accepting a risk of catastrophic loss in exchange for short-term profits. If I'm a financial officer who can gain a million dollar bonus by investing in risky derivatives, which incur a 5% risk of a hundred-million dollar loss for the company, I could last several years and walk away with millions before the "Black Swan" of the loss occurs. Each year, I could report massive profits, while nobody would be monitoring the risks that build up.

But Black Swans are not limited to derivatives investors; when you look closely, they appear throughout our society. For example, imagine being a worker in the World Trade Center on September 10, 2001, confidently planning your career and your future. Or, going further back in history, if you were a citizen of the Austro-Hungarian Empire in 1913, you might have assumed that your life and livelihood were quite secure, and any probable change would be incremental and easy to accommodate – until the Black Swan of World War I arrived. Also think about your own life and career; are you exactly where you envisioned being 10 or 20 years ago? Chances are that some Black Swan occurred at some point in your life, causing a dramatic shift in one direction or another. If a particular alumni recruiter had not scheduled an extra off-season trip to Carnegie Mellon 22 years ago, planning to recruit an intern in my exact research area, my life would be quite different today. I would barely have heard of Oregon, known Intel only due to a sticker on my PC, and the world might have even been deprived of *Math Mutation*.

Ultimately, Taleb's contention is that nearly all major changes in history and society, as well as in people's personal lives, have resulted from these Black Swans, unpredictable individual events that had low probability but hugely disproportionate effects. This contrasts with the tendency of history books to treat everything as a predictable flow, what Taleb calls the "narrative fallacy". We have a natural instinct to construct stories and explain events as rational and predictable – but the real test is to see who actually predicted them in advance. Your history book, for example, talks about how World War I was a natural consequence of various tensions building in Europe; but how many investors restructured their portfolios in 1913 to take account of the coming high-probability "inevitable" war? Surprisingly few. If it had really been so predictable, surely people would have bet on it with their own money.

So, if these low-probability but high-consequence events are really prime factors in reshaping our lives and societies, does that mean we are all adrift and at the mercy of randomness? Taleb's answer is no. We just need to recognize this factor, take into account our lack of knowledge of these upcoming Black Swans, and use it to guide our actions. For example, Taleb is a fan of free-market capitalism over central planning, but for different reasons than many libertarians: the free market allows us to 'roll the dice' on many different opportunities, until we are lucky enough for the Black Swan of a Ford, Intel, or Apple that comes across some set of innovations that vastly improves society. Central planners, cautiously trying one low-risk option at a time, almost never offer that level of innovation.

In your individual life, it is also important to take the Black Swan factor into account. You need to watch out for situations where you face a negative Black Swan: a common example we are all familiar with is to buy fire insurance, incurring a small expense now so that the Black Swan of a house fire will not wipe us out. And conversely, position yourself to take advantage of positive Black Swans: at some point in your life, you may encounter a rare opportunity with unusually high potential payoff, like that recruiter visit that shifted my direction in grad school. Be sure to take advantage of such a case. They will be rare, but chances are they will happen from time to time.

But most importantly of all, recognize that all the pontificating talking heads who "explain" every unusual event in history, in the news, or in your life are mostly full of hot air, unless they actually predicted these events in detail *before* they occurred. Chances are that they are engaging in the narrative fallacy, rationalizing low-probability events after they occurred in order to sound intelligent.

Comparative Disadvantage
From Math Mutation podcast 165

Have you heard about the Law of Comparative Advantage? This is a famous economic discovery of the 19th century which is counterintuitive at first, but makes a lot of sense once you start plugging in some basic numbers. Suppose one country is better at producing everything than another country. This law states that in many such situations, it's advantageous to everybody if each country produces what they are best at, and they trade with each other, rather than having the higher-producing country wall itself off from the other. This law forms one of the basic arguments for free trade, and is often misunderstood and derided by those who don't fully understand it. We should point out that, like many economic theories, the mathematical modeling it uses is simplified compared to the real world, and there are some legitimate criticisms of the theory that are a bit too detailed for this podcast. But Nassim Nicholas Taleb's Black Swan concept, which we discussed in the previous section, adds a surprising new twist to the law.

To start understanding Comparative Advantage, let's look at a simple example of this law in action. Suppose France can produce wine for 1 dollar a bottle, and beer for 2 dollars a bottle, while Germany can produce beer for 3 dollars a bottle, plus wine for 4 dollars a bottle. Initially, France is better at everything, so can wall itself off and refuse to trade with Germany. Here I'm talking purely in terms of the numbers, so we're ignoring the social unrest that would occur in a population forced to drink French beer. Anyway, let's also assume each country has 120 million dollars to spend on beverages, and wants to consume roughly equal amounts of wine and beer. Some quick algebra will tell you that France will produce 40 million bottles each of wine and beer, while Germany will produce a little over 17 million bottles of each, for total consumption of about 57 million bottles of each by the two countries.

Now, let's suppose they decide to trade with each other, and each country specializes in what it does best: so wine is produced for $1 per bottle in France, and beer for $3 in Germany. If the combined nations' 240 million dollars is used to produce equal amounts of wine and beer again, we can produce 60 million bottles each of wine and beer: so the total consumption of the two countries increases by 3 million bottles of each. Even if they are a little drunk from subsisting solely on a beer-wine economy, the French and Germans should both be able to see the advantage of trade in this situation. The key insight here is the concept of opportunity cost: previously for each bottle of beer they produced, France had to give up the opportunity of producing 2 bottles of wine. Once they are allowed to trade, they are saved from this opportunity cost.

So, what's the new twist that comes from Taleb's theory? Remember the concept of the Black Swan: we must account for low-probability but high-impact events that may be unlikely at any given time, but are probable enough that they will happen in the long run. Suppose France takes the law of comparative advantage to heart, and decides to devote almost all of their nation's efforts to producing wine. Then one year, a new grape blight appears in France, killing almost their whole crop. The nation will be in serious trouble. In general, Black Swans which destroy productivity for a time in a particular industry, from causes such as disease, natural disasters, or wars involving key suppliers, are going to happen at some point over the long term. This danger is not just theoretical – we can see examples in history such as the Irish Potato Famine, where Ireland was over-dependent on a single crop which failed nationwide due to disease. So even in the presence of comparative advantage, it's important for nations to diversify to some degree, rather than devoting their entire economy to a very small set of optimal industries.

I should point out that some web commentators have incorrectly grouped Taleb's critique with the many critiques of Comparative Advantage based on social or political values, whose merits or non-merits are beyond the scope of this book. I think these discussions are sometimes missing the point – the Black Swan idea is just as mathematical, in some sense, as the theory of comparative advantage itself. The fact that something that seems unlikely in the short term may be significantly likely in the long term is a basic element of probability theory. For example, if one of a group of low-probability disasters has a 1% chance of occurring in any given year, it might seem like a remote worry. But some elementary calculations, calculating n for which $(.99)^n < .5$, we see that in 69 years, a single human lifetime, such an event has over a 50% chance of occurring. And in real life, we can't make calculations this precise: we know there are many things that can go wrong, but we can never have enough data to fully model them quantitatively. So we need to defend against these unknown disasters that are likely to wipe us out in the long term by making sure our productive capacity and our set of suppliers remain diverse, rather than over-concentrated in a small set of industries, and ensuring that we are robust overall in the face of these Black Swan events.

Money for Math
From Math Mutation podcast 182

Recently the media has been full of stories about Yitang Zhang, a struggling math professor whose career has included stints as an accountant and as a *Subway* sandwich maker. Zhang made a major advance towards proving the Twin Primes conjecture. Twin Primes, as you may recall, are pairs of prime numbers that just differ by 2, like 3 and 5, or 41 and 43. He didn't actually prove the theorem that there are an infinite number of pairs of twin primes, but proved a related theorem: for some number N less than 70 million, there are infinitely many prime pairs separated by N. I'm not going to rehash that story in detail – if you're reading this book, I'm sure you've seen or read it somewhere anyway by now. But today's topic is inspired by an amusing question someone asked me after reading this article: "So, how much money does he get?"

First we should ask the question: why would there be a monetary reward for solving certain math problems? The perception that this should be the case is probably a result of the publicity surrounding the Clay Institute's Millennium Prize, a million dollar reward announced in 2000 for resolving any of 7 famous math problems. This institute was founded by a successful businessman named Landon T. Clay, a pretty standard way these things get set up. Sadly, the twin primes conjecture is not one of the 7 prize problems, so even if he had proven the full theorem, Zhang would not have won the million dollars. There are numerous other monetary awards given in mathematics as well, though nothing coming close to the Clay Institute. Zhang has won several of these awards, for a total of almost $150,000, so we don't need to feel too sorry for him.

Probably the most interesting story of a math award is the origin of the Wolfskehl Prize for Fermat's Last Theorem, awarded to Princeton's Andrew Wiles in 1996. As you probably recall, Fermat's Last Theorem is the famous 1637 conjecture that the equation $a^n + b^n = c^n$ has no whole number solutions for $n > 2$. For hundreds of years, until Wiles finally solved it, many mathematicians spent their lives working on proofs of this theorem. The award for the theorem was created by German mathematician Paul Wolfskehl. According to legend, Wolfskehl had been rejected by a woman and was very depressed, so he decided to commit suicide at midnight one night. But, since he made this decision in early evening, he had a lot of time to kill until midnight, so decided to hang out at the library and browse through books and journals. As he was reading an article by mathematician Ernst Kummer, he found a flaw in one of the proofs related to Fermat's Last Theorem, and began working on some calculations to enable him to fix the article. Ultimately he was able to fix the flaw in Kummer's work, but this took him all night, causing him to miss his scheduled midnight suicide. This work didn't prove the theorem, but at least he was able to improve an article refuting a failed path to a proof. By morning, he was so excited about his original contribution to mathematics that he no longer wanted to kill himself. In celebration of the theorem that saved his life, he modified his will to create a 100 thousand mark prize for anyone who did prove it.

As with many great stories, however, there are some Internet suggestions that it may not really be true. There are a few other legends going around about the Wolfskehl prize. One is that he started out intending to be a doctor, but due to multiple sclerosis was unable to practice that profession, and went into mathematics instead. According to this theory, the award was to celebrate the opportunity that math gives to the disabled. Another story says that the prize was due to spite: because of his disability, Wolfskehl was forced to marry the only woman who would take him, a mean, desperate woman who treated him horribly. So he left his money to create a mathematical prize instead of leaving it to her.

In any case, should people go into math in the hopes of growing rich through prize money? I think there are probably much easier ways – if you're good at math and money is your sole motivation, your odds of striking it rich are probably much better in the financial industry. In many cases, your odds might be better buying a lottery ticket, for that matter. It takes a special kind of genius and dedication to solve most famous unsolved math problems – look at how many years Zhang spent on the fringe of his profession before his success – and by the time Wiles won the Wolfskehl award, it was only worth about 45 thousand dollars, a few months' salary for him. And the Fields Medal, the 'Nobel Prize' of mathematics, is only worth a third of that. If you're one of those talented people on the track to solve one of these problems, I suspect that a big monetary prize is the least of your concerns.

Liking the Lottery
From Math Mutation podcast 191

If you're the kind of person who listens to math podcasts, you've probably heard the often-repeated statement that a government-run lottery is a "tax on stupidity", due to the fact that stupid people are likely to waste their money on something that has a negative expected value. But is the case really that open-and-shut? Does the negative expected value of a lottery automatically make it not worth playing, and would this situation reverse if the expected value ventured into the realm of the positive?

Let's start by reviewing the concept of expected value. At a basic level, this is the sum of the possible values of your lottery ticket, each multiplied by the probability of getting that value. As a simple example, suppose you have a local lottery where you pay 1 dollar to guess a number from 1 to 10, and if you're right, you get 6 dollars back. Your expected value is 9/10 times -1, since in 9 out of 10 cases you lose your dollar, and 1/10 times 5, for a total of -40 cents of expected value. This represents the likely average return per round if you play hundreds of rounds of this lottery. Expected value calculations for real-life lotteries are similar, except that they deal with very tiny probabilities and values in the millions. Real-life lotteries almost always have negative expected values; for example, one website recently calculated the expected value of a U.S. Powerball ticket at -$1.58. Actually, for real life lotteries there are some complicating factors that reduce it further: you have to account for possibly splitting the jackpot with someone else who guessed the same numbers, and also the hefty chunk of taxes that Uncle Sam will take out of your winnings, but let's simplify this discussion by ignoring those factors.

Now here's the critical question: suppose after many weeks of a growing Powerball jackpot, which happens sometimes if there is no winner, the pot grows to the hundreds of millions, and the expected value crosses over into positive range. Is it now a more rational decision to play the lottery? I would argue no: you are still much more likely to be struck by an asteroid or lightning, die from a bee sting, or suffer a plane crash than to win. The expected value calculation really only kicks in if you are buying millions of tickets, in which case you can use it to figure out if your massive bet is likely to be profitable. One article I found online talks about an investment group that actually did try to buy all the tickets to a Virginia lottery one year, but was a bit hosed by the fact that not all the tickets could be printed in time.

Another way to realize the limited usefulness of the expected value is to think about a slightly odd lottery, as suggested in a blog by statistician Alan Salzberg: suppose you could spend all your savings for a 1 in 1000 chance to win 10 billion dollars. If you have less than 10 million dollars in the bank, this game actually has a positive expected value. Would you play it? I think 99.9% of people would think playing such a game is insane. When you can only play it once, you need to think about things other than the statistical average of thousands of trials. What is the likely net effect on your life if you play it once? Chances are overwhelming that this lottery would leave you penniless.

So, if the expected value calculation doesn't make sense, how do we figure out if playing the lottery is rational? I think the key factor is the cost to you of spending the price of the lottery ticket. Assuming you are doing OK economically, spending a couple of dollars every week can probably be considered effectively zero cost: you are likely to casually spend more than that on potato chips from vending machines, lattes at Starbuck's, etc. For this near-zero cost, what value do you get? There is that thrill of scratching off the numbers or watching the drawing, and having that infinitesimal but nonzero chance of becoming an instant millionaire; given the low cost, maybe that alone makes it worthwhile. You also know that your ticket cost has contributed operating funds to your government; your feelings on that may vary, but even if you're a hardcore libertarian, there are likely at least a few government services you are OK with funding. And you're probably happier giving money voluntarily than being required to by law, which would be the effect if nobody played the lottery and income or sales taxes had to be raised instead.

Now I'm not saying that lotteries are always a good idea: the arguments I just made are predicated on the fact that the lottery ticket is effectively zero cost to you. If it is not – if the 2 dollars per week would make a real difference in your life, or you are spending more money than you can afford to lose – you need to realize that the chances of winning something are so infinitesimal that this is really not a wise expenditure. It's always kind of sad when I see blue-collar-looking people pump what seems like hundreds of quarters into Oregon Lottery video poker machines in bars. Saving up the money annually to buy asteroid insurance would be more likely to benefit their families in the long run.

But overall, does playing the lottery mean you're stupid? This looks to me like an area where many people blindly apply a mathematical formula without really thinking about what it means. Assuming the cost of a lottery ticket is effectively zero when compared to your income, it looks to me like the answer is no, playing the lottery may be perfectly rational.

Number Nonsense
From Math Mutation podcast 70

Recently I was at a local fast food restaurant, and I saw a burger that looked good for $1.98. I ordered two of the burgers, plus a soda which was 99 cents, and the cashier told me the total was 6.93. I looked at her, confused, and tried to reason with her. "You see that burger I ordered is about 2 dollars? And I got two of them, plus a one dollar soda. Shouldn't the total be around 5 dollars?" She was a bit annoyed. "So, you're not going to pay the amount displayed on the register?" I valiantly tried one more time to reason with her. "Look, $2 + 2 + 1$ equals 5. So the total should be close to 5 dollars. Something is wrong here." She gave up and went back to get the manager. I could overhear her speaking to him, though she didn't realize it. "An irate customer up front is refusing to pay for his meal." Needless to say, after the manager finally sorted things out, it turned out she had rung me up for an extra burger.

But I was still flabbergasted that a high-school-age American would lack the basic skills to estimate that $2 + 2 + 1 = 5$, and therefore something had gone wrong. My confusion was cleared up when I did a bit of internet research, and found that many school systems are using calculators from as early as the first grade level. Don't get me wrong, I think in some places, such as advanced science classes or application of formulas, it might make sense to use a calculator in school. But thinking back to my elementary days, I think I gained a lot of my inherent "number sense" from working out lots of simple calculations on paper. Sure, technically it's the same if you remember $2 + 3 = 5$ from repeating a lot of hand calculations, or if you type 2 and 3 into a magical black box and get 5 – but the thought process is a lot different. When you do it by hand, you can't help but notice patterns and gain an inherent 'feel' for the numbers. If all you do is mechanically type them in and always get a guaranteed answer, you may lose even your basic impulses of curiosity about what's happening.

On the net, there seems to be a clear backlash these days against the trend of using calculators in schools. Like me, many educated adults are horrified to see that if they ask for simple calculations like 300 divided by 3, a sizable portion of calculator-educated teenagers will reach for a calculator rather than thinking about the problem and giving a quick answer. No matter how sophisticated our technology gets, there will always be human fingers, or brain waves, supplying inputs to our computing devices – and thus always room for human error at some point. Having a basic number sense, and being able to understand whether simple, round arithmetic results are in the right ballpark, will always be a critical component needed to sanity-check calculations and avoid disasters. In a few years will we hear about a bridge collapsing because some entry in a computer was off by a factor of 10, and nobody noticed?

Does the House Always Win?
From Math Mutation podcast 140

If you're a self-respecting math geek, by now you've probably heard the urban legend of the group of MIT students who made millions in the 1990s playing blackjack at Las Vegas casinos, immortalized in Ben Mezrich's book *Bringing Down the House* and the movie "21". I read the book over the holidays, and really enjoyed it; not great literature, but a breezy and entertaining read. What I found most surprising was the basic simplicity of the card counting scheme that the MIT students used.

To review the basics, blackjack is a simple card game where you are dealt two cards, and can request additional ones to try to hit a total of 21 without going over. Face cards count as 10, and aces can be 1 or 11. Because the dealers hand out cards from the same deck for many hands in a row, it's a game with a "memory", where the previous hands do have a real influence on the current hand. Compare with something like the dice game craps, for example: there, every die roll is truly independent, so looking for some kind of pattern is hopeless, as the odds of each number on a die roll are constant. In blackjack, it's possible in theory to gain an advantage by remembering the cards that have been dealt so far. Traditional counting techniques have been estimated to provide a player an edge of roughly 2 percent in the long term.

If you think about this 2 percent advantage, this means that if you play for a long time with a $100 average bet, you will win an average of $102 per hand. But this very slight advantage means that in the short run, you are likely to lose many hands, so you would need to be able to tolerate large losses and have the bankroll to come back. So unless your stakes are large and you play for a long time, there is not that much profit to be made. In addition, conventional wisdom said that this technique, "card counting", has been nearly impossible for decades due to the casinos' mixing 6 decks together, creating a huge number of cards to keep track of.

But the students on the MIT blackjack team realized that the 6-deck policy created a new vulnerability: if an imbalance ever developed, such as the remaining cards in the 6-deck stack containing a large concentration of face cards, knowledge of this situation could create a huge advantage for the players. And the advantage would last for a noticeable amount of time, due to the time it took to go through 6 decks' worth of cards.

The trick was that this kind of situation was not common enough that a single player sitting at a table would be able to use it. So they would send a group of "spotters", players looking for this situation, to multiple different blackjack tables at the casino, while a big better would wander around waiting for a signal.

The spotters would play continuously, placing minimum bets and keeping a running count, a single number that reflected how many more face cards than low cards remained in the deck. The number would start at 0, and each time they saw a low card (2-6), they would subtract 1 from the count, and each high card (10, A, or face card) would add 1. They also had to factor in the number of remaining decks in the shuffle, by dividing the count by that number: an imbalance during the first few hands would likely be very temporary, while one developed when halfway through the cards was likely to yield much bigger rewards. This counting wasn't as easy as it sounds. Multiple hands of cards are flying by rapidly at a casino, most dealt to other players, and to avoid suspicion of being a card counter they had to participate in random table banter, order drinks, flirt with the Vegas bimbos, and otherwise act like 'normal' gamblers. They did a lot of practice in mock games before risking this in a real casino.

When the imbalance in the remaining cards reached a certain point, where disproportionate number of face cards remaining would provide a distinct advantage, a spotter would cross their arms, signaling a "big better" to come in. This would be another team member, acting ostentatiously rich and throwing around money, who would sit down at the table without acknowledging that they knew the spotter and start making large bets. The spotter would "pass the count" to him, making nonsense remarks containing code words they had carefully memorized in advance. For example, suppose the current count was 9. They would say something like, "Darn, I forgot to feed my cat." The word 'cat' signaled the number 9, since a cat has 9 lives. The big better would then know the current level of imbalance, and choose the bets accordingly. By placing big bets at the right times, and only placing minimum bets when establishing the count, the team greatly magnified their potential return. This is how they won millions of dollars.

Once a casino realizes someone is using this technique though, it's pretty easy to fight. The simplest way is just to shuffle the cards more often rather than waiting until all 310 cards of the 6 decks are used, dramatically cutting down on the potential window of vulnerability created by any imbalance. Another easy method is to make a house rule not allowing new players to join a table or existing players to increase their bet except right after a shuffle. One might wonder why they don't just shuffle after every hand in the first place, completely eliminating the issue of the cards 'remembering' state that can be leveraged. Mezrich points out that the casinos actually *want* some level of counting to be theoretically possible. This way they can attract foolish gamblers who play blackjack thinking they have an advantage, when they actually lack either the counting skills or large bankroll needed to truly beat the house.

In the end, such fanciness wasn't really needed to stop the MIT team. The casinos figured out that this small group of people was consistently winning, and ended up passing around faxes with their photos, resulting in them being banned from most venues. Casinos have the power to ban suspected cheaters without needing any kind of proof. This is useful since legally, card counting is not even considered cheating, since it all happens in the heads of the players. If you believe Mezrich's book, there was further intrigue involving Hollywood makeup artists, armed thugs, crooked cops, and backroom beatings, though the *Wikipedia* article claims that these aspects of the story were largely fabricated. In any case, now that the book and movie are out, I doubt even the smallest casino will ever again be vulnerable to this exact technique. But those MIT guys are pretty smart – I wouldn't be surprised if even now, they have another team milking the casinos using newer and trickier techniques.

Mutating Up the Corporate Ladder
From Math Mutation podcast 146

If you're a regular listener to *Math Mutation*, you're used to hearing about the surprising corners of modern math, things like fractals, higher dimensions, and complexity theory. But you're probably convinced that, while they are fun to talk about, these strange ideas will not have much impact on your daily life. What if I were to tell you, though, that applying modern mathematics could provide you with useful leadership insights in day-to-day business situations? It sounds a bit counter-intuitive, but that's the contention of business author Margaret Wheatley. Let's take a look at a few of her ideas.

First, let's review the concept of fractals, like the Koch snowflake, which we discussed in earlier chapters. It has the nice property that if you magnify any piece of it by any amount, it seems to be a smaller version of part of the original snowflake: this endless self-similarity is a common characteristic of fractals. In the case of the ideal mathematical objects, however much you magnify such fractal shapes, you see additional complexity, all resulting from simple rules or equations used to generate the fractal. Real-life situations such as mountains and coastlines act like fractals to some degree, though of course their self-similarity only works to some finite level rather than infinity.

So, how do fractals provide insight into leadership? To start with, think about the set of values and policies advanced by your company. If your company is well-run, the set of corporate values visible in the top levels of the organization should match those visible at lower levels. In theory, you should be able to visit a front-line manager on the factory floor, and he should be promoting a corporate attitude matching that of an executive meeting held by the CEO. And this should arise naturally from solid basic principles, just as the simple rules for the Koch curve generate the complex snowflake: you should not be attempting to centrally specify every edge and boundary. If your company is not exhibiting this fractal-like self-symmetry, something has broken down at some level, and you're in a dangerous situation where the corporate goals are not being well-propagated. You need to understand where the symmetry ends in order to correct the situation. As Wheatley says, "The very best organizations have a fractal quality to them."

Another aspect of fractals that Wheatley finds relevant to business is their infinite complexity. When you take a 'big picture' look at a fractal, you see the illusion of a definite boundary. But upon a closer look, you see more and more complexity at the edge, making it impossible to properly measure this boundary. As Wheatley states, "Fractals suggest the futility of searching for ever finer measures of discrete systems." You can probably see how this would apply to business: managers love to specify quantitative measures of how their organizations are doing. If you work at a big company, I'm pretty sure you have been frustrated at some point by a top-down requirement to measure your work using some arbitrary statistic. We have to recognize that there are a huge number of measurements you can take, and in the quest for the perfect precise measurement, you can easily get lost in an endless refinement of measurements that can never finish. You have to step back and look for some underlying structure of the big picture, rather than focusing on the details. Just as you need to try to find the simple set of rules that generated a fractal to understand it, you need to think about true root causes deep in the organization rather than just measuring output at the edge to understand your business.

Wheatley applies many more ideas from modern math and physics in her book, *Leadership and the New Science*, and her many articles published afterwards. Naturally, there is some question about whether she is truly using these ideas to influence business thought, or merely invoking them with strained metaphors to sound impressive as she applies common-sense business advice. I'm a bit suspicious, for example, of her leaps from the difficulty of measuring fractals to the need for qualitative intuition: the generating function of a fractal is just as mathematically precise an operation as measuring the lengths of curves. I get especially skeptical when she starts to quote from questionable authorities like New Age quantum physics abuser Deepak Chopra or psychic dog researcher Rupert Sheldrake. To some extent, her ideas seem ripe for the kind of debunking Alan Sokal does of misused math and science, as we discussed is our chapter on Science and Skepticism. On the other hand, as I see it, nearly every business book I've read has essentially been about puffing up common-sense ideas to sound like new insights. If bringing in a metaphor from higher math helps to convince your local managers to implement more sensible policies, why not use it? And if babbling about cool stuff you heard in *Math Mutation* can provide you a leg up next time you're being interviewed for a job or promotion, why not take advantage? Be sure to send me a cut after your mathematical jargon wins you that huge raise.

The Converse of a CEO
From Math Mutation podcast 205

Ever since I was a small child, I aspired to grow up to become a great Rectangle. When I was only six years old, my father took me to meet one of the leading Rectangles of New Jersey, and I will always remember his advice: "Be sure to have four sides and four angles." All through my teenage years, I worked on developing my four sides and four angles, as I read similar advice in numerous glossy magazines aimed at Rectangle fans. In high school, my guidance counselor showed me many nice pamphlets with profiles of famous Rectangles who had ridden their four sides and four angles to success. Finally, soon after I turned 18, I took a shot at realizing my dream, lining up many hours to audition for a spot on the popular TV show *American Rectangle*. But when I made it up onto the stage, I was mortified to be met by a chorus of laughter, and end up as one of the foolish dorks that Simon Cowell makes fun of on the failed auditions episode. With all my years of effort, I had not become a Rectangle, but a mere Trapezoid.

OK, that anecdote might be slightly absurd, but think for a moment about the premise. Suppose you want to become successful in some difficult profession or task. A natural inclination is to find others who have succeeded at that, and ask them for advice. If you find something that a large proportion of those successful people claim to have done, then you conclude that following those actions will lead you to success. Most of us don't actually aspire to become geometric shapes, but you can probably think of many miscellaneous pieces of advice you have heard in this area: practicing many hours, waking up early every day, choosing an appropriate college major, etc. I started reflecting on this concept after looking at a nice career planning tool aimed at high school students, which lets them select professions they are interested in, and then read about attributes and advice from those successful in it.

Unfortunately, this kind of advice-seeking from the successful is actually acting out a basic mathematical fallacy. In simple logic terms, an implication statement "A implies B", is logically different from its converse, "B implies A". Neither statement logically follows from the other: "A implies B" does not mean that "B implies A". When we look at the case of rectangles, this seems fairly easy to understand: the condition A of having four sides and four angles does NOT imply the consequent B, that the object is a rectangle. By observing that all rectangles have these characteristics, we are learning the opposite: Being a rectangle implies that the object has four sides and four angles. This is important to recognize because there may be infinitely many non-rectangle objects that meet this condition, and actual rectangles might represent only a small portion of the possibilities. If we wanted to isolate conditions that will imply something is a rectangle, we need to look at both rectangles and non-rectangles, to identify unique rectangle conditions, such as having four right angles. Once we have a set of properties that will pertain only to rectangles and not to non-rectangles, then we might be able to come up with an intelligent set of preconditions.

Sadly, real life does not always offer us geometric shapes. When we substitute a real aspiration people might have, too many try to infer the keys to success just from looking at the successful. Without thinking through this basic logical fallacy about a statement and its converse, "A implies B" does not mean "B implies A", many people waste lots of time and money following paths where their likelihood of success is minimal. A common case among today's generation of middle-class kids is the hopeful young writer who decides to major in English. An aspiring writer might see that many successful writers have degrees in English, without taking the time to note that the proportion of English majors who become successful writers is infinitesimally small. The statement "If you are a successful writer today, you probably have a college degree in English" does not imply "if you earn a degree in English, you will probably become a successful writer." In contrast, if looking at computer engineering, they might see a similar profile among the most successful – but will also find that unlike in English, a huge majority of computer engineering majors do end up with a well-paying job in that field upon graduation. So in that case, the implication really does work both ways – but this is a coincidence, since the statement and its converse are independent.

Even famous business consultants are subject to this fallacy. Have you heard of the influential 1980s business book *In Search of Excellence*, where the authors closely looked at a set of successful companies to find out what characteristics they were built upon? That became one of the all-time best-selling business books, and many leaders followed their sweeping conclusions, hoping to someday make their companies as successful as NCR, Wang, or Data General. But some have criticized the basic premise of this research for this same basic flaw: trying to determine the conditions of success by looking only at the successful will inherently get you the wrong kind of implication. It may enable you to find a set of preconditions that being successful means you must have had, while these same preconditions are met by endless numbers of failed companies. You really need to study both success and failure to find conditions that uniquely imply success.

So, when you or your children are thinking about their future, look carefully at all the available information, not just at instances of success. Always keep in mind that a logical statement "A implies B" is truly distinct from its converse "B implies A", and take this into account in your decision making.

CHAPTER 11

■ ■ ■

Looking at Life

It's often the case that math and physics geeks look down on the "squishy sciences" like biology and medicine. However, mathematical foundations are critical to these sciences: it was largely the transition from qualitative observation to careful, quantitative measurement that led to our amazing modern success in these areas. In addition, spending many hours watching *Animal Planet* with my daughter, I especially enjoy stories where animals or plants seem to have somehow discovered laws of mathematics, or made use of ideas from these areas in the struggle for survival.

Aside from this, you may have heard of the promising research area of Artificial Life, where scientists try to model idealized animal behaviors with computer programs: this can often provide useful insights into both actual animal behaviors, and into new ways to use those behaviors in creating more intelligent computer programs. You may have seen the lifelike behavior of John Conway's famous "Game of Life" simulation online; if not, I strongly encourage you to do a web search for the topic. It's described to some degree in the articles below, but no text description can do it justice; from a few simple mathematical rules, you can construct what seems to be a living simulation.

Hence, topics related to medicine, biology, and animal behavior have been a regular source of ideas for *Math Mutation*, and here I share some of my favorite episodes that touch on these areas.

Florence Nightingale, Math Geek
From Math Mutation podcast 85

You probably recognize the name of Florence Nightingale, nicknamed "the lady with the lamp", who is known as the founder of modern nursing. But most likely you only have a vague idea what that means in practice: did she find a way to give shots less painfully? Or discover a new method of changing bedpans? Actually, during the 19th century, she introduced modern concepts of cleanliness and hygiene into British hospitals, saving countless lives by preventing many common infectious diseases that were essentially untreatable in her day. But one lesser-known aspect of this achievement was that Nightingale's successes would not have been possible without her love of mathematics and her introduction of statistical methods into nursing and public health.

© Erik Seligman 2016
E. Seligman, *Math Mutation Classics*, DOI 10.1007/978-1-4842-1892-1_11

As the daughter of an upper-class British family, Nightingale was expected to master the social and home arts, marry a respectable partner, and participate in high society. But from a young age she insisted on being allowed to study math, and after many emotional battles her parents finally gave in, even hiring the famous number theorist J.J. Sylvester as one of her tutors. She took the subject to heart, even sprinkling occasional math references in her personal letters. Once after attending a political lecture, she wrote to her boyfriend, "I have invented a new system of Logarithms (finding the capacities of arithmetic not sufficiently extensive) to count the number of times 'Imperial Majesty' occurs in the speech."

Nightingale shocked her family by choosing to pursue nursing, and running off to care for soldiers at the Scutari Hospital in Turkey during the Crimean War. Once she arrived at the hospital, she was horrified at the filthy conditions, and began taking a detailed accounting of her activities as she defied the local military authorities and began to clean it up. For example, during one week she recorded the removal of 215 handcarts of debris, 19 flushes of the sewer system, and the animal carcass removal of two horses, four dogs, and a cow.

More importantly, though, she began to keep comparative statistics measuring hospitals against each other. She recorded that Scutari had 300–500 square feet per patient, versus 1600 in London hospitals. Furthermore, she recorded the death rates: 42.7% before the cleanup of Scutari, and 2.2% afterwards. When the authorities were unconvinced that this was more than a fluke, she found two control groups to compare to: patients in ordinary London hospitals who were typically much less seriously injured than soldiers at Scutari, and soldiers treated in the field who were too badly wounded to transport. She successfully collected and analyzed data showing that before her cleanup of the hospital, soldiers were better off being treated in the field – but afterwards, Scutari had a better survival rate than the best London institutions. To help persuade others of the importance of her measurements, Nightingale also invented a new type of pie graph, known as the 'polar area diagram', essentially a pie graph that shows quantities through the increase or decrease in the radius of the circle at a given angle.

After her achievements at Scutari, proven through her detailed measurements and statistics, Nightingale went on to continue a distinguished nursing and statistical career. Among her later achievements were advising the Union army in the U.S. Civil War, the founding of a women's medical college, and the introduction of modern medical care and nursing to India. She received awards including the Royal Red Cross and the Order of Merit, was elected an honorary member of the American Statistical Association, and was the first female member of the British Royal Statistical Society.

The Genius Who Cheated
From Math Mutation podcast 112

I'm sure that in high school biology class, you learned the story of the Austrian monk Gregor Mendel, who meticulously cross-bred many varieties of peas in his garden in the 1850s and 1860s, and came up with the first modern theory of genetics. By carefully observing a set of traits of the plants in his garden, he concluded that various traits were determined by inherited factors, later called genes, and that every plant carried two of these genes for each feature, one from each parent. He was ahead of his time, and his work was largely ignored until the 1900s, despite being a major step forward for our understanding of biology. But did you know that according to many 20th century statisticians, he almost certainly fudged his results?

To understand why it's likely he cheated, let's start with a look at his theories. As one example, he took two stable lines of plants, one with purple flowers and one with white, and cross-bred them. In the first generation, all the descendants had purple flowers. But by cross-breeding those, in the next generation he got 75 % purple and 25 % white. This perplexing result could be explained by the fact that each plant in the first generation inherited a dominant purple gene and a recessive white gene. Then, when two from that generation were bred, there would be a 25 % chance that the offspring would have two purple genes, and thus have purple flowers; a 50 % chance that the offspring would get a purple and a white gene, also having purple flowers due to the purple dominance; and a 25 % chance of two white genes, resulting in white flowers.

Mendel did this type of analysis based on seven different characteristics of his plants, and the result was his theory of inheritance, with two major laws. The law of segregation said that when an individual reproduces, it passes on one of each gene pair to its offspring. The law of independent assortment stated that genes mixed independently during reproduction. To some approximation, these theories were correct, and were a major step forward in our understanding of genetics.

But there were several respects in which Mendel was very lucky. First, in real life, the law of independent assortment isn't quite right: if two genes are near each other on the same chromosome, they will be very likely to be passed on together. Mendel happened to choose characteristics that were largely independent, not being near each other on any chromosome. Another complication is that for many characteristics, there is not complete dominance of one trait over another, but a combination: for example, if he had been dealing with snapdragon flowers, he might have found that white and red parents blend to a pink offspring. A third issue is that many real-life characteristics have multiple genetic contributors: the short-haired 'Devon Rex' and 'Cornish Rex' cat breeds look similar, for example, but their short hair is caused by two different genes. Mendel's lucky choices of characteristics to analyze eliminated all these difficulties.

Most damning, however, is an analysis done by statistician R.A. Fisher in the 1930s. He used what is known as a 'Chi-square' test. This is a statistical technique where you sum up squares of the differences between expected frequency and observed frequency of an event, divided by the expected frequency. This sounds odd, but has the nice property that it generates a graph whose area to the right of the observed value is proportional to the "P-value", or probability that the result occurred by chance. Using this technique, Fisher determined that Mendel's results were too good: there was only a 4 in 100,000 chance that his results would so closely confirm his hypotheses. In other words, the random combination of genes should have resulted in frequencies of plant traits that still confirmed Mendel's hypotheses, but not by such exact numbers.

What do we conclude about all this? Perhaps Mendel was a little too open about his theories to junior monks who helped tend his garden, and they were too eager to please their superior. Modern scientists are well aware that the best experiments are "double-blind", managed by assistants who don't know the intended hypothesis, to avoid the issue where even with good intentions, humans tend to subconsciously bias results. On the other hand, maybe Mendel just cherry-picked, or should we say pea-picked, his results, discarding bad trials as probably due to human error. Maybe we are just seeing what's known in modern times as the 'publication bias', where Mendel only talked about the subset of his experiments that seemed to give positive results, and discarded others as uninteresting or as the result of human error. Maybe he just was incredibly lucky. Some academics have even produced complex analyses that claim that for subtle reasons not taken into account by Fisher, Mendel's results are statistically fine as-is, though I found these a bit too difficult to follow.

In the final analysis, was Mendel a genius or a fraud? Despite questions about the validity of his experiments, he did develop foundational theories of modern genetics which are critical to current biological and medical research. So perhaps he was a bit of both.

Shuttle Butt
From Math Mutation podcast 72

Have you heard about how the width of the solid rocket boosters on the space shuttle can be directly traced to the width of an ancient Roman horse's rear end? Here's how the story goes.

Apparently the solid rocket boosters for the shuttle are transported by rail, and need to be correlated with the standard railroad gauge, or track width, in order to fit through tunnels. This standard width is about 4 feet 8.5 inches. This width comes from standards set in England. The English set this standard because early trains were designed based on similar horse-drawn wagons, and that was a standard spacing for wagon wheels. The wagon wheels have that spacing because they had been built for centuries to correspond to ruts in old roads. And the major roads in England and Europe were first built by the ancient Romans, who had a standard wheel width on their war chariots. And finally, the chariots were designed to match the width of the back ends of a pair of warhorses. Thus, the solid rocket boosters on the space shuttle were directly based on the width of a Roman horse's behind!

It's a cute story, but sounded a bit suspicious, so I checked out a few urban legend websites, like Snopes and The Straight Dope. My suspicions were confirmed – there are quite a few holes in this story. To start with, railroad tunnels are significantly wider than the tracks, though perhaps some tunnel-related limitation reduces the size of the shuttle's boosters. As for the standard gauge – there were a lot of creative American designers who experimented with different railroad track widths in the 19th century. During the Civil War, the Confederacy had three different sizes of tracks, and that may have contributed to their loss. The North had maximized efficiency by using the most popular width as a standard. After the war, destroyed Southern infrastructure was rebuilt (with central organization) to match the North, and that's how the U.S. gained one standard track width.

But more fundamentally, I think this story is a classic example of how correlation does not mean causation. In other words, just because one event occurs before another, it doesn't mean that the first one is the cause. It's true that at the dawn of the railroad era wagon wheels were about the same width as ancient Roman chariots – but that's because from ancient times through the Industrial Revolution, transportation technology was a constant. Whether an ancient Roman charioteer or a 19th century settler, you were using wheeled vehicles pulled by pairs of horses. So, the common cause of the width of a pair of horse's behinds explains the widths of ancient Roman chariots, medieval English wagons, and the first railroads.

Booms and Busts
From Math Mutation podcasts 14 and 159

According to the State of Missouri's website on rabbit management, each female can produce 35 young per season, and *Wikipedia* says rabbits live 5–10 years. If we make the conservative estimate that a pair of rabbits has 3 good reproductive years and rounding off to get an even 100 offspring, that means that each pair of rabbits results in about 50 times as many rabbits in the next generation. That means after 2 generations, there will be 50×50, or 2500, rabbits, and in general after n generations, 50^n, an exponential function.

For fun, now let's calculate how long it would take for the planet earth to become nothing but a gigantic ball of rabbits. The volume of the earth is about 1.08×10^{21} cubic meters. A cube about 8 cm on a side seems like a good estimate for the size of a rabbit, which would give a volume of .000512 cubic meters. So our planet is equivalent to about 2.10×10^{24} rabbits. Solving the equation $50^n = 2.10 \times 10^{24}$, by taking the log of both sides and then solving for n, we get an answer between 14 and 15. So, in other words, after only 15 generations of rabbits, there should be no non-rabbit matter left on our planet. That's quite an Easter.

This may lead you to an obvious question: why haven't the rabbits taken over? Well, fortunately, there are plenty of limiting factors, such as food supply, predators, etc, that prevent them from sustaining the ideal reproduction rate in this simplistic model. But when rabbits are placed in a new environment with lots of food and few predators, we do see explosive growth – just ask the people of Australia, who got 24 rabbits imported in 1859, and by 1869 could hunt two million annually without making a dent in the population. Yet given these calculations, it still seems somewhat surprising that *Math Mutation* has a nontrivial number of non-rabbit listeners.

Let's look into how we can model the countervailing forces, like scarcity of food or space. There is in fact a nice relation, originally developed by Belgian mathematician Pierre Verhulst in the 19th century, that models modified exponential growth in the presence of limited resources. He was actually describing a continuous model and using calculus, but when discussing populations we can use the simplified discrete form. Assume there is some maximum population or value that can be supported by the environment. Let $p1$ be the proportion of this maximum tolerance we have currently reached, r be the rate of growth, and $p2$ be the expected proportion of the next generation. Then the "logistic difference equation" can be expressed as:

$$p2 = r \times p1 \times (1 - p1)$$

In other words, we are simply multiplying an additional factor to the exponential growth: $1 - p1$. With regular exponential growth, we would just have $p2 = r \times p1$. The additional factor accounts for the fact that as $p1$ gets close to 1, the maximum possible size of the population, this will act as a tempering force to reduce the exponential growth. When we are very far from saturation, that $1 - p1$ is about 1, so it looks like standard exponential growth. The closer we get to saturation, the more the growth is throttled by that extra factor.

This formula doesn't seem very complex, but if we plug in numbers some very interesting things happen. Intuitively, how would you expect this equation to act? A good guess might be that if you start with a low value for p, population growth will initially be rapid, and then will smooth out and achieve some kind of stable equilibrium. For a growth rate r of 2.9, this does indeed happen: for example, if we start with a p value of .2, indicating we have about 20 % of our maximum number of rabbits, we can see it initially jump to .46, then .72, bounce back a little to .58, and eventually settle down to a consistent .66. This shows that our population will settle down at about 66 % of capacity.

But for other growth rates, we see very different results. There are many growth rates for which the population never does settle. As one example, if we use $r = 3.3$, we can see that after a few generations, the population starts oscillating between .48 and .82. In other words, we have generations containing 48 % of our maximum rabbit capacity, followed by generations at 82 %, going back and forth forever. For various other growth rates, the population ends up oscillating between 3 or 4 values. There are even many values, such as $r = 3.57$, where the value never settles down or oscillates – it just seems to jump around forever between arbitrary totals.

What does all this mean? It means that this simple difference equation is an example of a *chaotic function*. That is, a function where a slight change in initial conditions can result in widely varying results. In this case, we see that small changes in the rate of growth can give us a nice stable population, boom and bust cycles, or random fluctuations. You might expect that a slight change in the rate would create a corresponding change in the equilibrium value, but this simply isn't true: you may modify the rate a little and find you have lost equilibrium entirely, changed a former equilibrium into a new boom and bust cycle, or in effect randomized the results overall.

But seeing chaos come out of such a simple model, I think we've also discovered a surprising fact: in many parts of real life, boom and bust cycles may be a natural and expected occurrence. We tend to anticipate that things we encounter on a daily basis will reach some kind of equilibrium, whether it is the rabbit population in your yard, the national economy, the planet's climate, your weight, or your spouse's mood. When we see things not in equilibrium, we usually consider it a serious problem, and try to figure out what has gone wrong. After seeing the behavior of the logistic difference equation, though, we see that boom and bust cycles are something that really does arise out of simple, natural processes. A slight tweak to your inputs will not always get you the nice, predictable result you want.

So, next time you see world events or your daily life going out of control, take a step back before stressing about it. Maybe it is a real crisis, maybe it's totally beyond your control, or maybe you just need to modify parameters of a few of your real-life logistic difference equations.

The Boids and Bees of Leadership
From Math Mutation podcast 160

Recently my daughter was staring out the car window at a large flock of birds passing by, and asked me, "How do they know which way to go?" As often happens with her questions, I didn't have a good answer, so I just said "They're birds, they are born knowing." She pouted a bit, instinctively recognizing that that was not a very good answer. Thankfully, I was saved by her short attention span. But as luck would have it, that same week I started reading an entertaining book called *The Perfect Swarm* by Len Fisher, which describes many types of group algorithms and intelligence, including the simple methods that many animals use to keep their position in a flock. The key is to recognize that no master planning is needed for birds to form a flock: the flock emerges from a small set of local rules used by each individual bird.

A classic demonstration of how simple individual rules can emerge into a group flock behavior was a computer simulation known as *Boids*, created by Craig Reynolds back in 1986. He created a computer animation of a group of triangular birds (remember, this was 1986, cut him some slack on bird shapes) that would fly around the computer screen in a realistic-looking flock. But rather than having a complex central program to plan out the flights of all the boids in his flock, he gave them each just three basic rules: Separation, Alignment, and Cohesion. The Separation rule required each boid to try to keep a reasonable distance from neighbors and avoid crowding. The Alignment rule required each boid to steer in the average direction of its closest neighbors. And the Cohesion rule required the boid to try to steer towards the average position of its neighbors.

Amazingly, these basic rules were all that was needed to create an accurate-looking simulation of real life flocks or swarms of animals. And when supplemented with an obstacle avoidance rule, this created cool effects like the flock automatically splitting and re-forming when blocked by a large tree-like obstacle. This influential demonstration has since been used as the basis for many realistic swarm animations in movies and computer games, including the penguin swarm in *Batman Returns* and a bunch of flying aliens in *Half-Life*.

But the uses of boids are not limited to entertainment. Scientists have since used them as a starting point and tried to nudge Boids-like simulations closer to behaviors of real-life animals, leading to surprising new insights. One classic case was a couple of biologists who decided they wanted to find out how swarms of bees are led to newly discovered flowers. You probably can vaguely remember from biology class that the bees dance to tell each other of discoveries. But if you think about it, the tight spaces in hives and the massive numbers of bees mean that in practice, only a tiny proportion of a swarm can actually see the dance and know where a newly discovered nectar source is. So how do huge groups of bees arrive in the right place? At first you might think that the scout and the few who saw the dance would somehow be designated as leaders and get their companions to the right place, but you would be wrong.

After many painstaking hours of observing actual bee behavior, the biologists discovered the bees' secret: they do NOT have designated leaders, and none of them really know who is going the right way. Most of the bees are just massed in a boids-like swarm, following local rules to fly with their neighbors. But the small number of bees who know the right direction, or the "streaker bees", bias their flight to go faster than average and in the right direction. These bees are distributed throughout the swarm: they may be near the front, middle, or back. Since they are going slightly faster, they eventually reach the front and have to veer back within the swarm for a moment. But with most of the bees just trying to follow the basic rules and stay in the swarm, the few who know where they are going are good enough: with just a small number who are flying quickly in a purposeful direction, the entire group will end up in the right place. The biologists confirmed their discovery with a Boids-like simulation.

So, in other words, with a huge mass just trying to stay with their neighbors, all it takes are a few who are moving purposefully to get the entire swarm in the right direction. Some have pointed out that this insight can apply in other areas of life. How many times have you been with a bunch of friends deciding what podcast to listen to, and since most of them had no strong preference, your enthusiasm for *Math Mutation* soon turned into a group consensus? Maybe that's not a realistic example, since after all, who could imagine dissent on that topic. But you can easily think of how this might apply to business: the visionaries in a company can "lead from within", and if everyone else is basically carrying out standard tasks through momentum, the few who are moving purposefully can steer the whole company in a new direction.

I should probably end this section here before I get into trouble: I think I just gave my boss an excuse not to promote me, since I've just proven I don't need to be artificially placed at the head of the flock to lead the team in a new direction.

A New Kind of Decade
From Math Mutation podcast 166

Recently Stephen Wolfram's massive tome, *A New Kind of Science*, referred to as NKS for short, celebrated its 10th anniversary. Normally, a title like that is a dead giveaway that the author is a crackpot, as I'm sure you'll hear if you listen to one of the dozens of skeptical podcasts that seem so popular on iTunes. But Wolfram is a special case – he is beyond doubt a true genius, having made contributions to particle physics starting at the age of 15, received a Ph.D from Caltech at the age of 20, was the youngest-ever recipient of a MacArthur "genius" grant at 21, and was the primary developer of the *Mathematica* software package, which has become a worldwide standard for computational math. On the other hand, being an actual genius doesn't prevent someone from going a little nuts after a while. So, did Wolfram, as the title of his book suggests, really revolutionize science? Or has he pulled a Linus Pauling and gone off the deep end?

First, let's review the core concepts of this New Kind of Science. The NKS concept is based on the idea that very simple computational models can lead to surprising complexity. One example is Conway's Game of Life, a member of the class of simple systems known as cellular automata. This "game" is played on an endless grid of squares, on which each square can be in a live (black) state, or dead (white) state. Each turn, any live square with exactly 2 or 3 live neighbors stays alive, otherwise dying of loneliness or crowding. Any dead square stays dead, unless it has exactly 3 live neighbors, in which case it becomes alive. That's it. Despite the simplicity of these rules, there is an amazing variety of behavior it can produce, depending on the initial configuration of live squares. On the *Wikipedia* page you can see a "glider gun", a structure of live squares that launches an endless series of "gliders" off to infinity. It's even been shown that the Game of Life is computationally universal: a general Turing Machine, and thus any real computer can be simulated with the proper configuration of starting squares on a Life grid.

Given that computational universality is possible in such a simple system, what other complex real-life phenomena can be modeled by cellular automata or similar systems? Such systems have been used as models for fluid dynamics, traffic, crystal growth, and various areas of geology and ecology. If you play with the Game of Life long enough, you will find some configurations seem almost literally life-like in the way they defy prediction and seem to have minds of their own: "artificial life" has even branched off as a separate area of study. Wolfram's contention seems to be that all the complex areas of modern science, such as biology, chemistry, quantum theory, and relativity, will eventually be revealed to be computations in some simple cellular automata-like system. But since it's very difficult to back-solve and guess such laws from observing the system, we should be creating a new scientific discipline, where we systematically examine the classes of simple computational systems, like cellular automata, and the computations that they can perform. Studying these computations will eventually lead us to discovering the underlying principles at the core of the other sciences: hence, if Wolfram is right, NKS really is indeed a new kind of science.

So, 10 years later, did Wolfram's ideas really revolutionize science? Wolfram himself seems to think so: his anniversary blog entry points out that "there's now an average of one paper a day citing the NKS book", and points out that NKS models have been created for diverse areas including hair color in mice, clustering of red blood cells, patterns of worm appendages (yes, I also have to wonder about that one!), financial trading systems, and shapes of galaxies. He also points out that papers have appeared about almost all of what he characterizes as the 256 simplest cellular automata, showing that at least some are taking to heart his suggestion of focusing study on these simple systems. Since journal papers tend to reflect the real efforts of working science to some degree, it seems fair to claim that NKS has influenced a nontrivial amount of research.

On the other hand, some have argued that NKS was not really a new kind of science, but a logical progression of science in general. Scientists and engineers have always been looking for simple models of real-life phenomena: in a sense, this is the essence of scientific inquiry. Wolfram and other cellular automata researchers have made some important points, that simple models can enable emergence of much more complex-seeming phenomena than one would initially expect. But is it really the case that scientists in the varied areas Wolfram cites would not have found their simple models without the contributions of NKS? And while cellular-automata-based models have been successful in numerous niche areas, I think we're still waiting for the description of some major, general area of science, such as particle physics or relativity, in terms of such a simple model. If there was such a success, I think it would be all over the front pages of the popular science literature, and as far as I can tell, almost all of the NKS impacts that Wolfram cites are limited to articles in academic journals. Not to knock journal articles, which are a standard part of the scientific process, but when you claim to be revolutionizing science, I think the bar is a little higher. We also can't forget that many scientists may have felt obligated to reference NKS in any discussion that vaguely related to cellular automata, due to its status as a well-known and massive tome on the subject: one colleague confessed to me that this was actually the case when he cited NKS, even though his paper would have been nearly identical without ever hearing of Wolfram's book.

So, did Stephen Wolfram really create a New Kind of Science, or was he just hypnotized by the cool animations of cellular automata on his screen into making the concepts sound much more grandiose than they deserve? It looks to me like the jury is still out: assuming his literature analysis is accurate, I think it's safe to say that a lot of working scientists, engineers, and mathematicians are taking his ideas seriously, though we have yet to see the revolutionary breakthroughs that were to be promised by his title. But if anyone deserves the benefit of the doubt despite choosing a book title that raises the crackpot-detecting flags of every skeptic in the scientific world, it's Stephen Wolfram.

Basic Bugs
From Math Mutation podcast 167

After I released the podcast episode referenced in the previous section, a listener, Andrew from Switzerland, sent me an email with a fascinating link. This link was to an article on "Braitenburg Vehicles", a set of simple, abstract vehicles defined by psychology and cybernetics researcher Valentino Braitenburg in 1986, which show how very simple robotic designs can lead to apparent intelligence and emotions, as viewed by a naive observer. While the robots can be physically built, and have been built or simulated at several links you'll see in the show notes, Braitenburg really intended them as thought experiments, to better understand the basics of psychology. As he states in his book, "Get used to a way of thinking in which the hardware of the realization of an idea is much less important than the idea itself."

The initial Braitenburg Vehicle can be thought of as a small toy car, with motors for left and right wheels, and basic sensors on the left and right sides of its front that can detect levels of some preset element of the environment, such as light or temperature, and operate the motors accordingly. The simplest vehicle can be thought of as one that just has temperature sensors that are together connected to both wheels, so it moves quickly when in a warm area, and slows down or stops in a cold area. If you saw one of these things in nature, would you realize how simple its design is, or would you attribute to it some conscious desire to get out of the sun and cool off? Here is Braitenburg's answer: "Imagine, now, what you would think if you saw such a vehicle swimming in a pond. It is restless, you would say, and does not like warm water. But it is quite stupid, since it is not able to turn back to the nice cold spot it overshot in its restlessness. Anyway, you would say, it is ALIVE, since you never saw a particle of dead matter move around quite like that." Braitenburg may be stretching things a little, but I don't think his analysis is so farfetched, especially if you envision the vehicle dressed up to look more like an insect than a robot.

As you make the vehicles a little more complex, you might seem to be simulating bugs with slightly more complex actions and emotions. For example, think about the case where the left and right wheels can operate independently, with each sensor connected to the wheel on the same side. If the vehicle detects heat on one side, that wheel will spin faster, and it will turn away; it will appear that the vehicle is afraid of heat, as with our original example, but fleeing in more interesting ways as its independent wheels produce curved movement. If each sensor is connected to the opposite wheel, however, we will actually produce a very different behavior: it will head towards a source of heat, eventually ramming into it if possible. So our vehicles will look very similar to animals displaying attitudes of fear or aggression.

The next variant Breitenburg proposes is reversing the sign of the sensors' influence, so instead of causing wheels to speed up, activating the sensors causes the otherwise-moving wheels to slow down. Now let's look at the two variants we previously discussed, with each sensor operating the wheel on the same side or opposite side. If on the same side, the vehicle will eventually come to rest facing the heat source, since the wheel on the opposite side will move faster than the one on the same side. If on the opposite side, the vehicle will eventually come to rest facing away from the source, for similar reasons, and will be more vulnerable to getting started again due to the appearance of remote sources.

So you could think of the first vehicle as a Lover, that really wants to be near the heat source, and the second as an Explorer, who likes the heat source but is ready to turn away for other opportunities.

It's fun to think about what other slight variants will lead to behavior simulating different primitive emotions. Braitenburg wrote a whole book on the topic, where he goes on to add new twists such as multiple sensors, or "threshold sensors" that act discontinuously when certain values are reached in the environment, gradually building up into primitive neural nets and elements that seem to be leading to basic computers. These result in illusions of intention, likes and dislikes, memory, and other seemingly higher-level animal behaviors. He even goes on to discuss how Darwinian evolution could be used to create even more advanced vehicles.

The complexity resulting from these simple vehicles is fundamentally different from the cellular automata discussed in the last section in several ways, though. Coming from the domain of psychology more than automata theory, they are much less precisely described. These vehicles are not limited to as simple a set of operations as a typical cellular automaton, and seem to be open to eventually incorporating arbitrarily complex computing elements. Still, even before the major complexities are added, it is quite surprising how easy it is to mimic some common real behavior patterns we might see among animals in nature, just by tying together a few dumb machine components in an obvious way. Both Braitenburg vehicles and cellular automata can also be viewed as basic contributions the fringe discipline known as Artificial Life, the serious study of artificially-created lifelike systems as the real foundations for generating non-biological life forms.

Anyway, next time you are frustrated or feel guilty about squishing a cockroach you see scurrying around your kitchen floor, remember that its behavior may well be represented by a trivial Braitenbug vehicle, and that it may be a dumb biological machine rather than an evil genius conspiring to destroy your kitchen.

Bugged by Math
From Math Mutation podcast 179

Since she was in first grade, my daughter has loved to play with insects and other bugs she finds outside. I try to encourage any science-type interest, though I've had to give her some stern talks about taking this hobby into the house. Thus we watch a lot of *Animal Planet* shows together. Recently one show mentioned 17-year cicadas, a type of insect whose larvae emerge from the ground into adulthood every 17 years. But as my daughter wondered out loud, how did they come up with the number 17? Why would an insect choose this particular number?

A little web searching revealed a few answers. But to start with, we need to figure out what question we're asking. 17 years is an interesting life cycle for more than one reason. On the one hand, it's pretty long to start with – why would an insect derive an advantage by spending so much time between generations? Then, as an avid math podcast fan, you've probably noticed another interesting aspect of 17: it's a prime number, divisible only by itself and 1. Do the cicadas benefit from both long generation time and prime-ness?

According to one theory, the long generations are useful in cases where the climate is in a state of rapid change. Periodical cicadas evolved about 1.8 million years ago, at a time of climate instability. In these situations, there would sometimes be several years in a row of inhospitable temperatures, either too warm or too cold. You can easily imagine that if bad years come in clusters, waiting a while to return after a year in which you barely survive can give you a bit of an edge, and prevent a Dust Bowl from wiping out your species. Of course, depending on the actual temperature patterns, the long generation time could be good or bad, but one modeling study linked in the show notes showed that in the climate of that time, a 17-year cicada would have a 96% chance of its descendants surviving for 1500 years, while a 7-year variety would only have an 8% chance. That's assuming the main threat is the climate, and is not accounting for hungry predators or pet-seeking 6-year-olds.

The fact that their period is prime is another interesting adaptation. The most popular theory, originally proposed by famous biologist Stephen Jay Gould, is related to the fact that cicadas avoid predators by a method called "predator satiation". This is a fancy way of saying they just reproduce in such enormous swarms that the predators can eat all they want, and there are still enough survivors to propagate the species. But there's a fundamental problem with this evolutionary strategy: predators could learn to increase their own population periodically in anticipation of this tasty swarm. If the period was a small, predictable number, it would not be too hard for some predator by chance to evolve a matching cycle. But with the prime number 17, chances are very small that a predator would evolve through random mutation to increase its population by the right period.

Another competing, or perhaps complementary, explanation comes from the cicada version of racial segregation: the subspecies that remain most distinct are the ones least likely to interbreed. For example, a 13-year and a 17-year cicada species would only meet at most every $13 \times 17 = 221$ years, meaning opportunities for interbreeding would be rare and the two genetic lines would be more likely to remain separate. There does seem to be a basic flaw in this explanation, if you think about it a bit: a 17-year period seems like an overly-complex way to keep species separate. All it would take is for a pair of species to have, for example, 2-year periods, but emerge 1 year apart – then they would never meet. However, if there are large numbers of random cicada groups of competing periods, it could be that among all these groups, the large prime-breeders will have the advantage of remaining distinct species. The smaller and non-prime breeders would meet much more often, and most likely blend with each other until they all seem to merge together.

As with many aspects of evolutionary science, it's kind of tricky to figure out empirically whether one or all of these explanations is the right one. But there are lots of links on the web describing computational models that support one or more of these hypotheses. Maybe one day you will figure out an even better answer. Personally, I'm happy as long as they don't end up in my kitchen.

Voyages Through Animalspace
From Math Mutation podcast 194

One day recently I was at the Oregon Zoo with my daughter, and we saw lots of cute and not-so-cute animals, including a tortoise, lizards, tigers, sea otters, and a chimpanzee. It's always amazed me that such a variety of animals could evolve on our planet, and through a variety of mutations some primal forms have led to all these diverse and dissimilar creatures. For a long time I found this hard to grasp, until I read Richard Dawkins' famous book *The Ancestor's Tale*. In one chapter of the book, he described evolution as a grand mathematical journey through a special kind of multidimensional space. Somehow this geometric view of evolution made it seem more real, and more sensible to me than it had ever been before, so I thought I would go ahead and share it with you.

What do we mean by a journey through a geometric space? Let's start by talking about a journey through an ordinary three-dimensional space. Think of a 3-D graph you might set up in a tool like Excel, showing your location in your house in terms of length, width, and height relative to the front door. So a dot at coordinates (0,0,0) might indicate that you are at the front door, while (10,10,10) might show that you are in your computer room a short distance away on the second floor. Suppose you ask the question: is it possible to get from the front door to the computer room? The answer is yes if you can draw a continuous path in your graph from coordinates (0,0,0) to (10,10,10), in which every point along the way is physically reachable. If your computer room is unreachable – say your wife encased it in steel walls on all sides to keep you from playing so many video games – this is represented by impassable blacked-out regions in your graph, preventing you from drawing this continuous path.

Now let's look at how we can model animal evolution as a graph. Think about several characteristics of animals, such as fuzziness, size, and strength. You could draw points on a 3-D graph showing where some similar animals fall in these dimensions. Perhaps your house cat would be close to the graph's origin, while a Siberian tiger would be represented by a point further out. Let's ask the question: can we travel on a continuous path, where motion is due to genetic mutations resulting in a living creature slightly different on one of our three dimensions, from the house cat point to the tiger point? It's pretty easy to imagine mutations that make an animal slightly larger, stronger, or fuzzier. Nobody would seriously propose, for example, that there is a blacked-out region somewhere between the cat and tiger where, after a certain size, there is no way a creature with that specification could be alive. So we can easily imagine that the cat and the tiger are related.

Looking at just these three dimensions is obviously a massive simplification, as there are thousands of dimensions along which an animal can be described: diet type, eyesight, hearing, and many other things you probably can't even conceive of if you're not a professional veterinarian or biologist. So the three dimensions we are limited to in our sad dimensionally poor existence, at least from our perception, are not sufficient to describe a creature. But the core concept remains: any animal can be thought of as a point in a large multidimensional graph. Graphs with more than three dimensions can be easily modeled with modern computer systems, though we can't physically look at more than three in a single figure. If you want to figure out if some animal can have an evolutionary relationship to another animal, you just need to ask: can you conceive a continual path from one to the other in this gigantic space? It doesn't matter if the path is incredibly long – evolution has millions of years to work with.

The most challenging part is that there are lots of blacked-out regions on this graph, representing non-viable monstrosities: the point with the size of a house cat and the bite strength of a tiger, for example, can probably never be reached, though if you try to pet my cat Manny while he's washing himself you may get pretty close. As Dawkins points out, "in the multidimensional landscape of all possible animals, living creatures are islands of viability separated from other islands by gigantic oceans of grotesque deformity. Starting from any one island, you can evolve away one step at a time, here inching out a leg, there shaving the tip of a horn, or darkening a feather." So the islands that Dawkins describes in his graph can be thought of as connected by thick sandbars, showing paths from one to the other where the intermediate creatures are reasonable. The journey from a T. Rex to a chicken may seem incredible, but I don't find it that hard to imagine a very long continuous series of changes that trace this journey in this strange type of space: changes in size, gradual transformation of arms to wings, hardening of teeth into a beak, etc.

There's actually one more detail of this space that makes evolution slightly easier to believe than it might sound at first. We've been talking about continuous paths, but that is an oversimplification. Every genetic change is actually a tiny discrete 'jump' from one point to another, so the paths do not have to be fully continuous. So, for example, the jump from total blindness to light-sensitive spots, then to recessed spots filled with fluid, and so on to a full eye may seem to have many discontinuities, but that's okay, as long as none of the discontinuities is large enough that it can't be jumped by a small genetic mutation. In other words, some thin blacked-out regions of this graph may not be insurmountable. There are of course some discontinuities that can't be jumped – a bird with a petroleum-based jet propulsion system might be plottable here, but it would require such a massive set of changes at once that it's probably effectively impossible. The blacked-out regions between our superbird and the chicken are likely just too thick to allow an evolutionary jump.

Anyway, maybe this only helps for math geeks, but I found Dawkins's spatial explanation a really intuitive way to think about evolution. Next time you play with your cat or dog, remember that he's not just a pet, he's a unique point in a massive multidimensional space.

A Heap of Seagulls
From Math Mutation podcast 201

The Heap Paradox, also known to snootier intellectuals as the Sorites Paradox (Sorites being the Greek word for heap), goes like this. We all agree we can recognize the concept of a heap of sand: if we see a heap, we can look at the pile of sand and say "that's a heap!". We all agree that removing one grain of sand from a heap does not make it a non-heap, so we can easily remove one grain, knowing we still have a heap. But if we keep doing this for thousands of iterations, eventually we will be down to 1 grain of sand. Is that a heap? I think we would agree the answer is no. But how did we get from a situation of having a heap to having a non-heap, when each step consisted of an operation that preserved heap-ness?

One reason this paradox is so interesting is that it apples to a lot of real-life situations. We can come up with a similar paradox if describing a tall person, and continually subtracting inches. Subtracting a single inch from a tall person would not make him non-tall, would it? But if we do it repeatedly, at some point he has to get short, before disappearing altogether. Similarly, we can take away a dollar from Bill Gates without endangering his status of "rich", but there must be some level where if enough people (probably antitrust lawyers) do it enough times, he would no longer be rich. We can do the same thing with pretty much any adjective that admits some ambiguity in the boundaries of its definition.

Surprisingly, the idea of clearly defining animal species is also subject to this paradox, as Richard Dawkins has pointed out. We tend to think of animal species as discrete and clearly divided, but that's just not the case. The best example from the animal kingdom may be the concept of "Ring Species". These are species of animals that consist of a number of neighboring populations, forming a ring. At one point on the ring are two seemingly distinct species. But if you start at one of them, it can interbreed with a neighbor to its right, and that neighbor can interbreed with the next, and so on... until it reaches all the way around, forming a continuous set of interbreeding pairs between the two distinct species.

For example, in Great Britain there are two species of herring gulls, the European and the Lesser Black-Backed, which do not interbreed. But the European Herring Gull can breed with the American Herring Gull to its west, which can breed with the East Siberian Herring Gull, whose western members can breed with the Heuglin's Gull, which can breed with the Lesser Black-Backed Gull, which was seemingly a distinct species from the European gull we started with. So, are we discussing several distinct gull species, or is this just a huge heap of gulls of one species? It's a paradox.

Getting back to the core heap concept, there are a number of classical resolutions to the dilemma. The most obvious is to just label an arbitrary boundary: for example, 500 grains of sand or more is a heap, and anything fewer is a non-heap. This seems a bit unsatisfying though. A more complicated version of this method mentioned on the *Wikipedia* page is known as "Hysteresis", allowing an asymmetric variation in the definition, kind of like how your home air conditioner works. When subtracting from the heap, it may lose its heapness at a threshold like 500. But when adding grains, it doesn't gain the heap property again until it has 700. I'm not convinced this version adds much philosophically though, unless your energy company is billing each time you redefine your heap.

A better method is to use multivalued logic, where we say that any pile has some degree of heapness which continuously varies: over some threshold it is 100%, then as we reduce the size the percentage of heapness gradually goes down, reaching 0 at one grain. A variant of this is to say that you must poll all the observers, and average their judgement of whether or not it's a heap, to decide whether your pile is worthy of the definition.

If you're a little more cynical, there is the nihilistic approach, where you basically un-ask the question: simply declare it out-of-bounds to discuss any concept that is not well-defined with clean boundaries. Thus, we would say the real problem is the use of the word "heap", which is not precise enough to admit philosophical discussion. There are also a couple of more involved philosophical resolutions discussed in online sources, which seem a bit technical to me.

Ultimately, this paradox is pointing out the problem of living in a world where we like things to have discrete definitions, always either having or not having a property we ascribe to it. It is almost always the case that there are shades of grey, that our clean, discrete points may reach each other by a continuous incremental path, and thus not be as distinct as we think.

CHAPTER 12

■ ■ ■

Puzzling Paradoxes

Paradoxes, statements that somehow seem both true and false or otherwise self-contradictory, have always been a favorite topic among math geeks. They seem to indicate cases where our desire for airtight logic and mathematical precision runs headlong into a brick wall. Often they show some kind of previously misunderstood limitation in the logical structures we have been building, or some kind of fundamental disconnect between our human languages and the underlying mathematics.

There are also some cases that seem like paradoxes at first due to limitations of our intuition of our linguistic description, but actually are perfectly consistent when correctly understood. I lump some of these into the same bucket as well, since they require stretching our minds in similar ways to properly dig ourselves out of the illusionary "paradoxical" hole.

You Can Cross the Road
From Math Mutation podcast 94

You have probably heard about Zeno's Paradox, a classical problem first described by the ancient Greek philosopher Zeno of Elea around 490–430 BC. More precisely, he described a series of paradoxes, but here we will discuss the most commonly known one, sometimes referred to as the Dichotomy paradox. It is equivalent to the following: Suppose you want to cross the street. Before you get all the way across, you have to get halfway across. But before you get halfway across, you have to get 1/4 of the way across. And before you get 1/4th of the way across, you have to get 1/8th of the way across. And so on. Since you have to do an infinite number of things before you advance any distance, it's impossible to get anywhere. Thus, all motion is an illusion.

Actually, to my mind, even if you take the paradox at face value, the conclusion that all motion is an illusion seems to me like it doesn't really solve the problem. After all, assume you want to imagine yourself crossing a road, since you are resigned to the fact that real-life motion is impossible. Before you imagine yourself getting all the way across, you must imagine yourself getting halfway across. And before you imagine yourself getting halfway across, you must imagine yourself getting a quarter of the way across. And so on. So, even in your imagination, it's a paradox, and an imaginary crossing of the street can't exist either! I guess if we can imagine ourselves teleporting across the road, that

E. Seligman, *Math Mutation Classics*, DOI 10.1007/978-1-4842-1892-1_12

would bypass this issue, but I'm pretty sure I can imagine myself crossing a road without teleportation. Of course, maybe I'm just imagining that I can imagine that.

Another interpretation of the ancients was that the paradox proves that space is not infinitely divisible. If at some point you can't take half the remaining distance anymore, there is no paradox. However, there are some difficulties with this solution: even if there is some size below which no physical objects exist, isn't it still the case that you can describe distances smaller than that, since nothing prevents you from halving an arbitrarily small number? And then you must traverse these distances in order to move. If we supplement this with the supposition that continuous motion does not exist, and below some distance motion is always achieved by micro-teleportation that we don't notice, I suppose that does resolve it. In fact, that interpretation may be consistent with modern quantum physics. But there are better ways to resolve the paradox and still allow continuous motion.

As early as the 300 s BC., Aristotle had noticed that the sequence of diminishing distances also comes with a sequence of diminishing *times*. Thus if it should take you a minute to cross the street, you should take 1/2 minute to get 1/2 way across, or 1/4 to get 1/4 of the way across, etc., so while you do have to do these infinite things, they begin to take infinitesimal amounts of time as they get smaller and smaller. While the total distance you need to cover is $1/2 + 1/4 + 1/8 + ...$, on to infinity, the total time is given by the same series. If each of the infinite distances took a full minute to traverse, then you truly would have a paradox: it would take you an infinite time to sum up the infinite terms and travel any distance. But since the infinite number of infinitesimal distances are covered using an infinite number of infinitesimal times, as long as these infinite series have finite sums, you are covering a finite amount of space in a finite amount of time, and there is no paradox.

By the way, remember that the infinite series $1/2 + 1/4 + 1/8 + ...$ does have a finite sum. To prove this, let S equal the sum of the series. If we write $S = 1/2 + 1/4 + 1/8 + ...$, then multiply both sides by 2, we get $2S = 1 + 1/2 + 1/4 + 1/8 + ...$ But the right side of that equation is just $1 + S$, since all but the initial 1 form the series we started with. And if $2S = 1 + S$, $S = 1$, and there is our finite sum. These kinds of results were actually known as early as Archimedes in the second century BC.

Now, some philosophers will say that this mathematical refutation based on summing infinite series is too glib, and in fact that the whole basis of allowing infinite processes in algebra and calculus is problematic. But I think calculus has worked in enough real-life contexts that I won't worry too much about it. Maybe you should listen to a philosophy podcast if you still have trouble imagining you can cross a road. But, as your hand will never be able to reach all the way to the keyboard to subscribe to that other podcast without summing an infinite series, that might be a challenge.

Four-Dimensional Greek Warships
From Math Mutation podcast 197

Today we're going to discuss the famous Ship of Theseus paradox. This paradox, known at least since the time of Socrates, involves the ship that the famous hero Theseus sailed from Crete to Athens. When the ship returned, the local shipwrights noticed that one of its boards was starting to rot, so replaced it. Because the ship was so famous, rather than eventually scrapping it as it got old, they continued over the years replacing any broken or decaying boards with new ones. Eventually 100 % of the wood on the ship had been replaced: not a single plank remained that had been present on the original voyage from Crete. At this point, was this still the same ship from Theseus's voyage, or should this be considered a new ship that had been constructed in the port of Athens? Many years later, Thomas Hobbes made this paradox even more confusing by adding one more issue: suppose someone had painstakingly gathered the removed boards and used them to construct a second, complete though not-very-sturdy, ship. Would this new ship have a better claim to be the true Ship of Theseus?

This paradox has been described with several variations over the years. During the enlightenment John Locke described essentially the same paradox based on patching a sock. Jules Verne came up with a clever version where an old man took a much younger wife, who after being widowed many years later took on a much younger husband, and the pattern repeated for hundreds of years. Was this "the same" marriage centuries down the line? In modern times, you're probably most likely to encounter this paradox in terms of rock bands, who seem to be in a perpetual state of warfare over the band name after some original members leave. For many years I was disappointed that the last *Velvet Underground* album, "Squeeze", was out of print; it was considered to not be a true VU album by most fans, since none of the original members were left by the time it was recorded, but I was still curious to hear it. Eventually *iTunes* made that album available, and after listening to it once, I have to take the side that the vessel in Athens' harbor should have been burnt as kindling.

But more seriously, how do we resolve this paradox? After all the planks had been replaced, is Theseus's ship the one made out of the new planks, the one constructed from the old planks, or neither? I think the most satisfying solution I have heard is based on the concept of "four-dimensionalism". The idea here is that our problem stems from the naive definition of Theseus's ship as an object at some point in three-dimensional space. We need to think of the Ship of Theseus as the union of a continuous set of objects in four-dimensional spacetime, accounting for not just the three dimensions of the physical object but also the points in the fourth dimension of time. Each 'slice' of the Ship of Theseus consists of a three-dimensional ship at a particular point in time, and the Ship of Theseus is a union of all these slices.

In this view, the Ship of Theseus consists of the original ship on the day it returned to port, plus the ship in that port with one plank replaced a week later, and so on. We need to be clear about what we are defining as Theseus's ship at each point in time. Thus gradually changing out the planks doesn't make the ship a different ship, since we defined the Ship of Theseus to be the one that is continually getting repaired over some interval in four-dimensional spacetime. Note that it also possible to instead define the Ship of Theseus to be the sum of the original planks at each given point in time. With this alternative but still valid definition, Theseus's oddly defined 'ship' will start out as the original ship upon its return from Crete, but then at many future time points consist of a partial ship plus a pile of wood in a junkyard somewhere, until Hobbes completes his duplicate.

Of course, there are simpler ways to resolve the paradox as well. One might argue that the real problem is just the vague and muddy definition of what is the Ship of Theseus, which the paradox assumes the user to have as an implicit notion but is never clearly stated. (I wonder if the "E-prime" techniques discussed a few chapters ago, where you avoid the word 'is' in favor of more specific verbs, could have helped here.) If we had clearly defined the Ship from the outset, in terms of its planks, its deed, the captain, or something more tangible, we would never hit a seeming paradox. A nice metaphor might come from some U.S. gun control laws, where the a component known as the "lower receiver" is considered to contain the identity of the firearm, and it's legally the same weapon as long as any other part is replaced, though it becomes a different one if you change the receiver.

Anyway, next time you bring your car to the local mechanic and they replace a tire or air filter, think about whether Theseus would let you tell your spouse that you brought home a new car.

A Christmas Surprise
From Math Mutation podcast 139

Suppose I tell you that I have a Christmas surprise for you: I will release a special *Math Mutation* episode some morning next week, but you won't know in advance when it will be. Being a math geek, you start reasoning about it logically.

Suppose the episode comes out on Saturday. By Friday, you would have known that the only remaining day of the week is Saturday, so you would already know in advance that the podcast will be released that day – contradicting my original promise that it would be a surprise. Once you eliminate Saturday, on which other days could it happen? Well, now you know it can't be Friday either, because if the podcast didn't arrive by Thursday, you would know it had to be Friday, since you already know it's not Saturday. If it's truly to be a surprise, it thus can't be on Friday either. Following a similar chain of logical induction, we see that it could not happen any day of the week! Then, when I actually release the podcast on Wednesday, you are truly surprised, as I originally promised – which seems to contradict your logic.

This paradox, most commonly known as the Surprise Paradox or Prediction Paradox, is said to have originated during World War II, when the Swedish government announced on the radio that there would be a civil defense exercise during the following week, but nobody would know the date in advance. Swedish mathematician Lennart Ekbom recognized this paradox inherent in the announcement, and it soon spread virally throughout the mathematical community. One of many reasons why civil authorities sometimes consider mathematicians a pain in the rear. Anyway, by 1948 the paradox appeared in print in a British magazine, and by the end of the 20th century, nearly 100 academic papers had appeared on the topic in philosophy and mathematics literature. There have been many equivalent formulations, including one involving a prisoner about to be hanged, and another by Martin Gardner where a husband promises a surprise gift to a wife.

This paradox does seem to be a valid issue when first described, but it's not too hard to satisfy yourself after thinking a few minutes that some kind of trick has been played on you. One simple resolution is to accept that I have truly given you a contradictory statement by promising that you can never know in advance the day of the podcast. If you can reach a logical contradiction by deductive reasoning, that means that one of your axioms, probably the statement I gave you, had to be false. Thus, my statements must not be truthful in the first place, and you actually have to disregard them, and have no useful information about the timing of the podcast.

A web author named Uri Geva points out another subtle aspect of this paradox. We are really mixing two forms of surprise here: there is surprise due to lack of knowledge, when you simply do not know when something will happen, and surprise due to contradictory knowledge, when you have logically reasoned that something will not happen and then it does. The announcement of the surprise podcast is transforming the first type of surprise into the second. Is it still a paradox? I think at its essence, this explanation is equivalent to the previous resolution, where the logical contradiction has effectively made the two forms of knowledge equivalent.

But to get a more satisfying resolution, it looks to me like the key point is the definition of 'surprise' in the context of the paradox. Am I truly promising you that you will never, at any point, have the possibility of advance knowledge of the correct day? Or am I merely promising that *today* you do not yet know the day of the podcast? If we use the former definition, there is likely a paradox here. If we use the latter definition, merely requiring that you don't know the day right now, then the paradoxical reasoning falls apart.

Given that the Surprise Paradox has inspired so many academic papers, you can easily imagine that a lot more has been said about it. A detailed *American Mathematical Monthly* paper by my old Princeton classmate Tim Chow formulates several versions of the paradox in symbolic logic, and even somehow calculates an exact probability distribution to choose the day while maximizing surprise potential for a five-day week. However, I think as long as you are clear on your definition of surprise, you can avoid a paradox without going into that much detail.

Resolving the Grandfather Paradox
From Math Mutation podcast 122

You have probably heard about the "Grandfather Paradox", a classic logical proof that backwards time travel is impossible. Suppose you have a time machine that allows you to travel back in time. You can then use it to go back to the time when your grandfather was a child, and shoot him. This would mean that you could never be born. But if you were never born, you would not exist to travel back in time and kill your grandfather – so your grandfather would be OK, and you would be born after all. Since our only logical conclusions are the contradictory results that you both were and weren't born, our premises must have been wrong, and time travel is impossible.

The most common way science fiction authors have resolved this issue is to use multiple universes: if you travel back in time and alter events, you create a parallel universe where the different events occur. So in one universe your grandfather was shot, and in another he wasn't. But that's a form of cheating: if traveling back in time puts you in another universe, then you didn't really travel back in time. Can we resolve this paradox in such a way that allows time travel in a single universe?

The answer is yes. Back in the 1980s, Dr. Igor Novikov developed something called the "Novikov self-consistency principle", which states that if any event would change the past in an inconsistent way, its probability is 0. So, you simply couldn't shoot your grandfather. You might try to do it – but something will always go wrong. Perhaps you will have a fatal heart attack a moment before firing the bullet. Or you might return to find your father telling the sad story about how when your grandfather was a child, some maniac from the future killed his lesser-known twin brother. Or you might discover you accidentally dropped your iPod in the past while time-traveling, and its inspirational math podcasts gave your injured grandfather the will to live. By the way, last year you may have read in the papers about some physicists trying to use this same self-consistency principle to explain why the Large Hadron Collider was plagued by problems – it would somehow lead into inconsistencies in the universe, so it was just not physically possible to get it working.

But how can this self-consistency principle make sense, in a world with free will? It might be useful to imagine it in terms of a two-dimensional example. Let's suppose we are in Flatland, the two-dimensional universe discussed in earlier chapters, and it is inhabited by a single being, Mr. Square. Think of the plane as a square sheet of glass, with Mr. Square being an etched square that can move around it. Let's take the plane as it exists at 1:00, then lay on top of it the plane at 1:01, then at 1:02, etc. As we pile on Mr. Square's plane-moments, each a single unmoving glass sheet, imagine we are building up a large translucent cube, created by all the stacked planes. Kind of like one of those etched-glass things you see in souvenir shops. We see a snake-like trajectory etched in the cube, showing Mr. Square's movements around the plane. The square is experiencing each moment in succession, his consciousness slowly moving up the cube – but we see it all at once. It's kind of a history-cube showing us in one glance the square's entire worldline, the history of all he has experienced or will experience. The etchings in the cube never move from our point of view, because what Mr. Square senses as his dimension of time has been translated into our dimension of height.

Now suppose Mr. Square has a time machine in his plane, and uses it to travel back in time from 2:00 to 1:00. If we look closely at our cube, we can see there is a second Mr. Square near the bottom, in the 1:00 plane. But no matter what trajectory the second Mr. Square traces, it can't change his original trajectory – there is only one cube, and we are seeing the world history all at once. Mr. Square might have the idea that he can travel back in time and kill his 1:00 self, but we can look at the cube and see that either the 1:00 Square continued to live or he didn't: the whole history-cube is static, and does not change from our point of view. So no matter what happens, Mr. Square's history must occur in a self-consistent way: it's physically impossible for it to occur otherwise. Similarly, what we perceive as time might be seen by creatures outside our universe as a mere static trajectory through an already-determined past and future, which by their nature cannot be changed in an inconsistent way.

This solution is a little disturbing, as it demolishes our perception of free will. But don't blame me – the universe destined me to write this chapter.

A Million Dollar Choice
From Math Mutation podcast 123

The previous section's topic, on resolving the Grandfather Paradox, got me thinking about another paradox that is somewhat related. This one, known as Newcomb's Paradox, has also led to large amounts of debate in the philosophical community, and also brings into question fundamental issues of the existence of free will. It goes like this.

Suppose we have a Predictor, that is, a psychic or similar person who claims to be able to predict the future, and has always predicted correctly in the past. The Predictor has set up two boxes, Box A and Box B, and given us the choice to take the contents of either Box B or both boxes. If he predicted that we would take both boxes, he has put 1000 dollars in Box A and nothing in Box B. If he predicted we would just take Box B, he has put 1000 dollars in Box A and a million dollars in Box B. It's important to note that the predictor set up the boxes in advance, so no action we take now can change the contents. Should we choose to take both boxes, and just Box B?

One obvious answer comes from the theory of "expected utility". If we take both boxes, that must mean that the Predictor expected us to do so, and we only get 1000 dollars. If we just take Box B, then the Predictor expected us to do this, and we get the million dollars. Of course, if we really believe in the Predictor, we don't really have a choice in the matter, but we can hopefully believe we are the kind of person with the willpower to just take one box, in which case we are about to be a millionaire. So to the extent that we believe we have free will, we should take Box B. Case closed.

Another obvious answer comes from the theory of "dominance". Since the money is already in the boxes, we are in one of two situations: either Box A has a thousand and Box B is empty, or Box A has a thousand and Box B has a million. In each of those cases, opening both boxes gets us $1000 more than just opening Box B. So, since opening both is guaranteed to get us more money than just one, we should naturally take both boxes. Regardless of what the Predictor thought we would do, the money is in the boxes, so it would be crazy to leave one behind. Case closed.

You can see why this is seen as a paradox – we have seemingly irrefutable arguments for taking just box B, or for taking both boxes. But I think the statement of the paradox cheated in a way: we were presented with a Predictor who knows the future, yet told we can choose between the boxes using our free will. Both can't be the case. Either the predictor's knowledge of the future is imperfect, or we don't really have the free will to make the choice. Since we start with contradicting assumptions, it's not surprising that we come up with contradicting conclusions. And a paradox is not a paradox if its initial premises cannot be consistently valid.

Another way to look at it is that we do have free will, but also backwards causality: that we can choose freely, and the choice we make now travels backwards in time to the Predictor's mind. But in this case, we are faced with something similar to the solution to the Grandfather Paradox discussed in the last section – as long as there is a way for the universe to be consistent, this kind of time travel is not ruled out. If such is the case, we clearly need to choose Box B, since that will cause the million dollars to be there. Newcomb specifically ruled this solution out, since the answer is too easy in that case.

By restating this paradox in ways where the Predictor is slightly imperfect, we can look at it in intriguing new ways. For example, suppose you learn that two contradictory genes run in your family. One will make you a football champion and lead to multimillion dollar NFL contracts. The other will make you the greatest math podcaster in the Internet universe, but will also cause you to die by the age of 30. You are entering college, and must choose whether to major in football studies or mathematics. You face a similar choice to the paradox: does it make sense to give up your lifelong dream of being a math podcaster, and settle for being a football champion, on the basis of the fact that choosing math means an early death is more likely? Or does it make more sense to reason that there's no way to change your genes, so you should follow your heart? Hopefully, you will never be faced with such a gut-wrenching choice.

The Painter's Paradox
From Math Mutation podcast 55

In earlier chapters, you may remember that we discussed an odd figure known as a fractal that had an infinite perimeter but a finite area. But in a recent discussion a co-worker of mine, Wayne, mentioned a simple non-fractal 3-D figure known as Gabriel's Horn that has a similar property: it has an infinite surface area, but only a finite volume! This means, bizarrely, that you can fill up the figure by pouring in a finite amount of paint, but you can never paint its surface. Thus the definition of this figure is sometimes referred to as the "Painter's Paradox". It was first discovered way back in the 17th century by Evangelista Torricelli, the same Italian physicist who invented the barometer.

How is Gabriel's Horn constructed? It's actually very simple. Take the graph of $y = 1/x$, drawn starting from $x = 1$ and continuing on to infinity. Now rotate the graph in a circle around the x-axis, forming what looks like a giant horn, with the wide end open at $x = 1$ and the narrow part trailing off forever to infinity. You can see an illustration of this figure below:

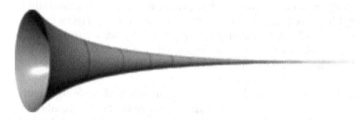

Figure 12-1. *Gabriel's horn*[1]

If you look at the circular cross-section at any given point, its radius is $1/x$, so its area is π/x^2. Now let's calculate the surface area and volume of this figure. You may recall that integrals, or infinite continuous summations of $1/x^2$, converge to a finite sum, since the values diminish very quickly and become negligible. So if we find the volume by taking the integral from one to infinity, we can easily determine that the total volume of this figure adds nicely to a value of π. But to find the total surface area, we need to take the integral of the perimeter, which is $2\pi r$ at any given point, or $2\pi/x$. So this surface area is proportional to an integral of $1/x$, which we know diverges to infinity. Thus, we could fill the horn with a finite amount of paint, but paradoxically, that paint could not fully cover the inner surface area. Actually, if we consider paint as a 3-dimensional object, and allow its thickness to diminish in proportion with the size of the horn, we can transform the painting into a volume problem and solve the paradox.

But something still seems seriously strange here. How can we have a finite volume, but an infinite surface area? Ultimately, in the particular case of Gabriel's Horn, this descends from our misfortune that sums of $1/x$ grow to infinity, while sums of $1/x^2$ converge – so paradoxically, adding a dimension to a figure whose size diminishes as $1/x$ will turn an infinite value into a finite value. Incidentally, you can see from this analysis

---

[1]Sourced from Wikimedia Commons at https://en.wikipedia.org/wiki/File:
GabrielHorn.png. Released into public domain by owner, user RokerHRO.

that the Horn is really just a representative of a huge family of similar figures, created by "stretching" an infinite two-dimensional graph into three dimensions using a converging function. For example, take the graph of $y = sin\ x$ from $x = 1$ to infinity, then just thicken it by stretching the graph in three dimensions by a factor of $1/x^2$ at each point, looking at the solid formed by the area between the graph's curve and the x-axis. You will similarly get a figure with infinite surface area and finite volume.

The real problem here is that we are defining figures that can never be physically built – you can't construct a horn that actually descends out to infinity. Any finite piece of Gabriel's Horn that can be built, can obviously be both painted and filled with paint. If we were describing a figure that could physically be constructed, I'd be a lot more worried about this paradox. But as it is, our description of the figure is really just an interesting mathematical game. You might find it fun to look up other converging and non-converging integral functions in a calculus text, and figure out more strange figures you can build by rotating, stretching, or otherwise extending the dimensions, creating out-of-sync area-related and volume-related properties.

The Monty Hall Paradox
From Math Mutation podcasts 3 and 109

The so-called "Monty Hall Paradox" is a famous problem in probability theory which a large majority of engineers and scientists will get wrong the first time they hear it. Recently I was stuck for several hours in a boring lecture, and after covering several notebook pages with scribbled diagrams, I think I finally understand the problem. Let's see if you believe me.

Suppose you are on a game show facing 3 curtains, which we'll call A, B, and C. You are told that one conceals a prize, and the others contain goats. You choose curtain A. But before it is opened, the host opens curtain B, revealing that there is a goat behind that one. Should you stick with curtain A, or switch to curtain C? You probably will reply with the common-sense answer that it makes no difference, and you have a 50/50 chance either way. But you are wrong. You should switch to curtain C, with 2/3 probability. To see this, think about it this way. There are three possible situations, all equally likely.

- Situation 1: You chose the correct curtain.

- Situation 2: You chose the wrong curtain, and the first of the ones you didn't choose contains the prize.

- Situation 3: You chose the wrong curtain, and the second of the ones you didn't choose contains the prize.

If you were in situation 1, where you chose correctly the first time, obviously you shouldn't switch. But in situations 2 or 3, you should switch. So with 2/3 probability, you will win if you switch curtains.

That explanation might sound somewhat satisfactory, but there is something about this problem that causes people to insist that it's wrong. We just have a really strong intuition that changing the door shouldn't matter. Don't feel too bad though: in one study, only 13 % of respondents got the answer right, and people who get it wrong include such respected intellects as scientists, Nobel Prize winners, and podcast hosts. Even the genius Hungarian mathematician Paul Erdős got it wrong the first time. When columnist Marilyn Vos Savant wrote an article on this problem in *Parade* magazine in 1990, she got about 10 thousand letters claiming her correct solution was wrong, including 1000 signed by PhDs. Eventually this resulted in a 1991 front page article in the *New York Times*, where with typical media savvy, they decided to settle the issue by interviewing former game show host Monty Hall himself. Surprisingly, he understood the problem quite well.

Here's another way to think about the solution: When you choose a door, you have a 1/3 chance that you chose the right one. In that case, you have to keep your original choice to win. But there's a 2/3 chance that you chose the wrong door – and in these cases, switching doors is what gets you the car. If you still doubt the explanation, you can find online descriptions of a procedure for simulating the game using playing cards, in addition to websites that run Java simulations, showing that indeed the probability is 2/3 that you win if you switch doors. But these explanations are still a bit unsatisfying: where does the odd result come from? Why isn't the probability 50-50?

I think the best insight comes from the fact that there is a hidden assumption here: the host knows which door has the car, and will always use his knowledge to open a door with a goat. The host's knowledge is what tilts the probability. If, on the other hand, the host does *not* know which door is the winning one, and has a chance of accidentally revealing the car, the whole game changes. Now there is a 1/3 chance that switching gives you the car, 1/3 chance that switching doesn't give you the car, AND a 1/3 chance that you chose the wrong door originally, but the host reveals the car & gives away the game! So, in this no-knowledge case, it is truly a 50-50 decision whether to switch or not.

Another question someone asked me is whether this Monty Hall paradox translates to the recently popular TV quiz show, *Deal or No Deal*. On this show, the contestant chooses one of 25 suitcases that contain various amounts of cash, and gradually opens ones he didn't choose. Often the choice eventually comes down to two suitcases, the one they chose and one they didn't choose, and the host asks them if they want to switch before the final suitcases are opened. In this case, we are in the no-knowledge situation, since the contestant chose which cases to open, with no influence from anyone who knows the contents. So it truly doesn't matter: the probability of winning is equal either way.

A Mathematical Nuclear Bomb
From Math Mutation podcast 24

In 1913, Alfred North Whitehead and Bertrand Russell completed the third volume of the *Principia Mathematica*, one of the most important works of the 20th century. This was a precise, detailed description of the foundations of mathematics, intended to enable the derivation of all possible mathematical truths from a set of well-defined axioms and inference rules. Mathematicians commonly skip a lot of steps when describing proofs, and understanding a work of mathematics usually requires a significant amount of background and intuition. But the *Principia* was intended to be fully self-contained, and not rely on such external knowledge: simply by using the axioms and following the rules stated in the book, all possible theorems should be eventually provable. To get an idea of the level of detail here, the proof that $1 + 1 = 2$ is not reached until page 379 of Volume I.

The first three volumes covered set theory, cardinal numbers, ordinal numbers, and real numbers. The authors had planned to continue with a fourth volume, but instead stopped due to intellectual exhaustion.

While the *Principia* was surely a monumental achievement, did they truly describe a consistent system that could eventually prove all truths of mathematics? For many years it was thought so – but in 1931, 25-year-old Austrian mathematician Kurt Gödel stunned the academic community by proving conclusively that there were true theorems that could be stated using the system in the *Principia*, but could never be proven. In itself, this might not be so bad; it might simply mean that a few more axioms needed to be added to complete the system. But Gödel's theorem went further than that. He proved that in principle, *any* consistent formal system complex enough to represent the natural numbers would have to contain true but unprovable theorems. Thus, no matter how many axioms Whitehead and Russell added, Gödel could always construct a new statement that we know is true, for reasons outside the system, but could never be proven from within.

The way he did this was by showing how to construct an almost-paradoxical statement, let's call it G, that essentially says 'Statement G can never be proven.' If you think about it for a moment, you will see why it follows that if the system is consistent, this is a true but unprovable statement. If it is false, then it means theorem G can be proven – but then we have proven a false theorem, and the system is inconsistent. On the other hand, if it is true, then that means we have created a true statement that we will never be able to prove. And this is not the only one. This is what is known as an "existence proof" – Gödel demonstrated the existence of true but unprovable statements, forever knocking down the notion that a complete formal system of mathematics is possible. Further results show there are actually an infinite number of such statements.

So what does this all mean? Well, it does not show that mathematics is useless; we still learn many powerful and subtle truths based on things we are able to prove based on our axioms. What it does show is that, in general, not every mathematical problem is solvable. Perhaps some of the Millennium Prize Problems will still be unresolved at the turn of the next millennium, simply because they describe true but unprovable statements. But in the meantime, I'm sure plenty of useful math will be done.

CHAPTER 13

■ ■ ■

Rethinking Reality

In this final chapter, we take a step back and question some fundamental notions of truth, falsehood, and our reality in general. Do we know what it means to make true statements, and consistently define our terms, or are we just playing word games? Is there a subtle connection between mathematics and religion? Is it possible to accurately reason about reality and existence? Do we exist at all?

Answering All Possible Questions
From Math Mutation podcast 207

Have you ever wished, in your daily life, that you had a simple way to find all the answers about any subject that was vexing you? Perhaps you are in a personal crisis wondering whether God exists, or maybe have a mundane issue as simple as finding your way home when lost. Well, according to 13th century monk Ramon Llull, you're in luck. Llull devised a unique philosophical system, based on combining a set of primitive concepts, that he believed would provide the path to solving any conceivable dilemma. His primary goal was to find a way to discuss religious issues and rationally convert heathens to Christianity, without relying on unprovable statements from the Bible or other holy books. As a philosophy, his system was far from definitive or complete, and gradually faded into obscurity. But along the way he became a major contributor to mathematics, making advances in areas as diverse as algebra, combinatorics, and computer science as he tried to elaborate upon his strange philosophical methods.

Llull began by listing a set of nine attributes in each of several categories of thought, intended to represent a complete description of that category, which could be agreed upon both by Christians and non-Christians. For example, his first list was the nine attributes of God: goodness, greatness, eternity, power, wisdom, will, virtue, truth, and glory. He wanted to discuss all combinations of these virtues, but repeating them endlessly was kind of tedious in the days before word processing, so he labeled each with a letter: B, C, D, E, F, G, H, I, K. He then drew a diagram in which he connected each letter to each of the others, forming a nine-pointed star with fully connected vertices. By examining a particular connection, you could spur a discussion of the relationship of two attributes of God: for example, by observing the connection between B and C, you could discuss how God's goodness is great, and how his greatness is good. Whatever you might think of his religious views, this was actually a major advance in algebra: while the

© Erik Seligman 2016
E. Seligman, *Math Mutation Classics*, DOI 10.1007/978-1-4842-1892-1_13

basics of algebra had existed by then, variables were commonly represented by short words rather than letters, and had been thought of as simply representing an unknown to be solved for in a single equation. For the first time, Llull was using letters to represent something more complex than numbers, and mixing and matching them in arbitrary expressions. In addition, his diagram of the relations between attributes was what we now call a graph, an important basic data structure in computer science. He also created another depiction of the possible combinations as a square half-matrix, another data structure that is common today but was unknown in Llull's time.

Figure 13-1. *One of Llull's wheels[1]*

Llull's system got even more complicated when he introduced additional sets of attributes, and tried to find more combinations. For example, another set of his concepts consisted of relationships: difference, concordance, contrariety, beginning, middle, end, majority, equality, minority. He also had a list of subjects: God, angel, heaven, man, imaginative, sensitive, vegetative, elementative, instrumentative. Even deeper philosophical conversations could theoretically result from combining elements from several lists. This created some challenges, however. He would again label each element of these lists with letters, but keeping track of all combinations led to an explosion of possibilities: just the three lists we have so far make $9 \times 9 \times 9$, or 729 combinations, and he had a total of 6 major lists. So to facilitate discussion of arbitrary combinations, he created a set of three nested wheels, each divided into 9 sectors, one for each letter. One would be drawn on a sheet of paper, and the other two would be progressively smaller and drawn

---

[1]Sourced from Wikimedia Commons at https://commons.wikimedia.org/wiki/File:Ramon_Llull_-_Ars_Magna_Tree_and_Fig_1.png, public domain under tag {{PD-Art}}.

on separate sheets that could be placed over the first one and independently rotated. Thus, he had developed a kind of primitive machine for elaborating the combinations of multiple sets: for each 9 turns of one wheel, you would turn the next larger wheel once, and by the time you returned to your starting point, you would have explored all the combinations possible on the three wheels. Several centuries later, the great mathematician Gottfried Leibniz cited Llull as a major influence when inventing the first mechanical calculating machines.

There were also several other contributions resulting from this work: Llull can be thought of as the first person to discuss ternary relations, or functions of more than one variable; and he anticipated some of Condorcet's contributions to election theory, which we discussed in our chapter on politics. Llull, however, was not really concerned with making contributions to mathematics, as he was concentrating on developing a comprehensive philosophical system. In his own mind, at least, he believed that he had succeeded: he claimed that "everything that exists is implied, and there is nothing that exists outside it". To help prove this point, he wrote a long treatise elaborating upon physical, conceptual, geometrical, cosmological, and social applications of his ideas. Apparently he even spent five pages showing how his system could aid the captain of a ship that was lost at sea. Personally, I would prefer to have a GPS. But even if our modern thought processes don't strictly follow Llull's guidelines, we still owe him a debt of gratitude for his contributions to mathematics along the way.

Sacrificing a Goat to Calculus
From Math Mutation podcast 65

The differences between mathematics and religion are pretty clear, right? In math, we only believe in clearly proven consequences of elementary assumptions. In religion, we must take things we don't really understand on faith. But is it always that way? In fact, when new mathematical ideas are initially being developed, we often understand them in a vague or imprecise form. Like in religion, mathematical speculation often begins with intuitive ideas about how something should be; rigorous definitions and proofs might be left until later. A classic example of this case is the initial definition of calculus. In the early 1700s, the philosopher Bishop Berkeley wrote a famous critique, claiming calculus was really a type of religion. And in some ways, he actually had a point.

To start with, let's review some of the basic concepts of calculus. You probably recall that a "derivative", or what Newton called a "fluxion", is the slope of a curve at a single infinitesimal point. But does this make sense? After all, ever since Euclid, we've known that a point can have lines through it at any angle, and there's no reason to prefer one over another. Just because the point happens to lie on a curve doesn't change this. Newton defined it by looking at the slopes of smaller and smaller lines crossing pairs of points on the curve, as you get closer to a single point – if they seem to converge to a known value, as is the case with most common geometric curves, than that can be said to be the slope at the point.

Bishop Berkeley's scathing critique of this method was called "The Analyst: A Discourse Addressed to an Infidel Mathematician". He pointed out that if you are assuming you have small intervals to obtain a slope, then compress those intervals to a single point, you have violated your initial premise – so you no longer can believe your information about the slope. He closes with a long list of final questions to ask the infidel mathematician, such as:

- Query 4: "Whether men may be properly said to proceed on a scientific method, without clearly conceiving the object they are conversant about, the end proposed, and the method by which it is pursued?"

- Query 16: "Whether certain maxims do not pass current among analysts which are shocking to good sense?"

- Query 63: "Whether such mathematicians as cry out against mysteries have ever examined their own principles?"

- Query 64: "Whether mathematicians, who are so delicate in religious points, are strictly scrupulous in their own science? Whether they do not submit to authority, take things upon trust and believe points inconceivable? Whether they have not their mysteries, and what is more, their repugnances and contradictions?"

It is questionable whether Berkeley was sincerely critiquing calculus at the time, or just trying to make a forceful case that if calculus were acceptable, we should provide more lenient criteria for considering Christianity. At the time, he was concerned about defending his religion against the Deists. And he was right that the concept of a limit, essential to a clear understanding of calculus's infinitesimal points, had not been clearly and rigorously defined – it was not until the 19th century that Bolzano, Cauchy and Weirstrass would precisely define limits using the concept of "epsilon-delta proofs". In a sense, scientists of the day were accepting something they had not fully defined or proven, in order to use Newton's fluxions in calculations.

However, Berkeley seems to have been glossing over the critical point that calculus provided experimentally verifiable results, and thus while not as rigorously sound as other branches of mathematics at the time, had a clear and demonstrated connection to reality. So his comparison with religion seems a little strained. Over the next few decades a number of books and pamphlets were published by Berkeley's contemporaries, refuting him point-by-point.

But this episode does serve to remind us that the "ideal" view of math we often see in school, where the edifice of proofs is slowly built up to precisely define everything we know today, is really just a small part of the story. Without dreamers taking intuitive leaps ahead of what they can really prove, humanity would never have achieved nearly as much in the mathematical arena.

Grue and Bleen
From Math Mutation podcast 76

My wife just looked out the window, and she thinks the grass on our lawn is green. But I know better. I told her that it's really bleen. Bleen is a color defined by 20th century philosopher Nelson Goodman, defined similarly to the following: An object is bleen if it appears green until July 31, 2018, and then blue afterwards. There is also a complementary color defined, called "grue", which describes objects that are blue until July 31, 2018, then green afterwards. These definitions may sound odd to you, but in my native land of New Jersey, things changed color from pollution all the time, so these definitions are perfectly natural. How can my wife prove that our lawn is green, and not bleen? Every day since we moved in, I've looked out our window, and it's looked bleen to me.

Being a clever reader, you might object – "Obviously your bleen and grue are complicated definitions, not natural notions of color like the green and blue that we are used to." But I can easily reply that you have it backwards. Earlier I stated the definitions in a form that seemed complex, but to me, bleen and grue are the natural notions. I define your bizarre color "green" as "bleen until July 31, 2018, and grue afterwards." So you see, it takes lots of words to construct this artificial color green that you are used to, while my simple notion of bleen can be stated in one word. Now, how sure are you that my lawn is not going to, in your terms, turn blue at some point?

This amusing and classic pseudo-paradox demonstrates the problem of inductive reasoning in the sciences. We observe natural phenomena, try to come up with the simplest explanations and definitions, and then use this information to model things mathematically. I'm pretty sure that at some fundamental level, blue and green are simpler than grue and bleen – perhaps the best answer I've heard to this conundrum is that blue and green can be defined without reference to time, by describing the wavelengths of light they represent, while grue and bleen cannot. But do we always get our inductions right? Is it always the case that the simplest explanation of phenomena we observe will turn out to be the most accurate?

One well-known example of a case we got wrong is Newton's laws of physics. His simple equations describing motion of common objects and forces seemed right for hundreds of years, until Einstein came along. Which do you think would be more surprising: my lawn turning blue on a certain day, or time slowing down a slight amount for me if I drive my car fast enough? Yet the latter case turns out to be true! What seems like a simple, obvious model of reality is not always so. We can never be sure that our inductive reasoning is right; our careful observations of past events always might be missing some crucial factor that would lead to a completely different theory. Take a look out your window – perhaps your lawn is bleen as well.

Hippos in my Basement
From Math Mutation podcast 113

I'm tired of being an engineer. I think I want to devote the rest of my life to the study of the wild hippopotamus. There are a few difficulties though. To observe hippos, I would have to fly to Africa, and I hate airplanes. I'm also not a big fan of hot weather. But using a basic principle of mathematics, I can work around these problems, and study hippos without leaving my basement. I'm not talking about cheating and looking things up on the internet: I'm talking about original scientific research on the wild hippo.

How do I accomplish this feat? Well, to start with, let's look at how ordinary science works. Often you have some proposition about the natural world, let's say for example, "All hippos have large noses". Think of this as a mathematical proposition "A implies B", where A is the property of being a hippo, and B is the property of having a large nose.

You would then go and do field research, observing large numbers of hippos, and determining what proportion do in fact have large noses. While you can't absolutely prove the proposition this way, you are gathering evidence inductively, which is usually the best scientists can do in many areas. Each confirming instance you observe adds weight to the hypothesis that A does imply B, and all hippos do truly have large noses.

Now let's suppose I want to study this theory, but make use of a little math first. Remember we started with the proposition that A implies B, where A is the property of being a hippo, and B is the property of having a large nose. But logically, A implies B is equivalent to (not B) implies (not A). Think about if for a minute, and I think you'll see why: if A implies B, then that means that if anything has the property (not B), it must not have (not A), otherwise the original implication would be false. For any statement "A implies B", "not B implies not A" is equivalent in truth value, and is known as the *contrapositive*.

So, instead of analyzing the original hypothesis that being a hippo implies having a large nose, let's look at its contrapositive: not having a large nose implies not being a hippo. Studying this is logically the same as studying the original theory. To verify this inductively, instead of having to look for hippos like I did in the original proposition, now I just have to look for things without large noses, and verify that they are not hippos. But that's easy! Just sitting here at my desk, I see my computer, my chair, my stereo, my cat, my daughter... Endless non-large-nosed things. And each one is a confirming instance of my theory, adding weight to my evidence that all non-large-nosed objects are non-hippos. In other words, without leaving my basement, I have been able to do original research into the theory that all hippos have large noses.

Now if you want to nitpick, you might point out that the number of hippos is so tiny compared to the number of non-large-nosed things, that if I spent my whole life cataloguing non-large-nosed things, I would still have gathered less convincing evidence than if I had gone to the zoo and looked at one hippo. Statistically, each piece of inductive "confirming evidence" using my method adds much less confidence to our hypothesis than would each observation of an actual hippo. But it doesn't change the fundamental logic of my argument. I'm still doing valid science in my basement, just doing it a little more slowly.

Is There a Hippopotamus in this Podcast?
From Math Mutation podcast 144

The discussion of hippos in the last section reminded me of a famous conversation between Bertrand Russell and Ludwig Wittgenstein, discussing the concept of truth and falsehood of logical propositions. Wittgenstein asserted that existential propositions are inherently meaningless, and as his retort, Russell asked him whether the statement "There is a hippopotamus in this room right now" is true or false. Wittgenstein held to his position, insisting that the question could not be answered, while Russell comically looked under all the desks for hippos. Various sources claim the discussion was about a rhino rather than a hippo, but I don't think the logic fundamentally changes.

The anecdote is amusing, and often used in popular discussions as a humorous jab at how philosophers are disconnected from reality. But as I thought about it, the statement bothered me a bit – after all, Russell was a leading mathematical thinker of the 20th century, and commented after the hippo conversation that he thought Wittgenstein was a genius. So there must have been some substance here, some sort of interpretation of logic in which Wittgenstein's obstinacy on the hippo statement actually did make sense. I began browsing the web for an explanation. I was surprised not to find any that I considered clear and concise: there were lots of references for humor value, or long philosophical treatises, but no straightforward answer to how this could be interpreted other than as a total disconnect from reality. Here's my attempt to summarize what I found.

To start with, there are the psychological explanations: there seems to be a persistent theme on the web of claims that Wittgenstein was likely suffering from Asperger's Syndrome, or a similar autism-spectrum disorder. One sufferer of this condition writes on the web: "Like Wittgenstein, we have a habit of hearing and seeing propositions, but feeling that they say nothing." As any of you who have met a philosopher know all too well, it is often tempting to explain their utterances as a result of a neurological condition. This would neatly tie up the question of Wittgenstein and Russell's conversation, but I don't find it very satisfying. It would certainly not explain Russell's respect for Wittgenstein's reasoning abilities. So I think we need to dig further.

One logical interpretation of the statement is that it stems from Wittgenstein's general belief that the concept of observation is invalid, and his refusal to admit the existence of anything except asserted propositions. Thus it would not be enough to look for a hippo using fallible human powers of observation: its existence would have to be argued based on fundamental propositions and definitions. If a proposition, such as the one about the hippo in the room, contained terms that had not been previously defined or could not be understood in terms of other propositions, it could not be labeled true or false at all, and would have to be dismissed merely as incoherent. Personally, I prefer to accept the evidence of my senses even if I haven't mathematically proven it, but I'm pretty sure I'm not as smart as Russell or Wittgenstein.

Another interpretation is that the statement depends a lot on the frame of reference and the thoughts of the speakers. For example, would a stuffed hippo Russell had in a desk drawer count? What about a picture of a hippo rapidly scribbled while Wittgenstein's head was turned? And how close are the two philosophers' ideas concerning the definition of a hippo? In a sense, this discussion could be about the limits of language to express thought and logic. Similar issues arise in another famous and controversial statement of Wittgenstein's: "If a lion could speak, we could not understand him." Speaking does not involve merely the communication of words, but their interpretation in terms of a conceptual scheme. A similar issue might arise from the title of this section: Is there a hippopotamus in this podcast? A lot depends on what I mean by the question. Does a hippo have to emerge from the page as I type this, or is the mere word enough? Since this section has been transferred from podcast to book form, is the question now nonsensical?

I should also point out that at least one source seems to imply that the commonly quoted version cuts off part of the sentence, and Russell's real statement was really "There is a hippopotamus in the room, but it cannot be seen or touched, heard or tasted or smelled". This throws a whole new light on things – this statement truly is one that cannot be checked through observation, and makes much more sense as something arguable by philosophers. It also makes everyone struggling to interpret the meaning of the shortened version kind of look like dorks, but since I'm one of them, I should probably stop now.

I Have Lied to You
From Math Mutation podcast 115

Suppose you accept the following statements: #1. If Statement S has a high probability of being true, you are justified in believing S. #2. If we are justified in believing S, and justified in believing S2, then we are justified in believing the statement "S and S2." #3. We are not justified in believing things we know to be false. These seem pretty reasonable, right? It's hard to imagine much of modern science proceeding without taking these as a basis.

But in 1961, philosopher Henry Kyberg spotted an apparent contradiction here, in what he called the "Lottery Paradox". Let's assume we have a fair lottery with 1000 tickets, and take our statement S as the statement "Ticket #1 will not win the lottery." This is 99.9 % likely to be true, so certainly we are justified in saying we believe S. Then let S2 be "Ticket #2 will not win the lottery." Similarly, we are justified in believing S2. Continue the process, and you will see we are justified in believing NO ticket will win the lottery. But we also know one ticket will win, leading to a contradiction.

In general, this paradox is seen as part of the philosophical field of epistemology, or the theory of knowledge. You can see that this paradox really applies to just about everything in our daily lives. How can you justify that you know you are reading this book? Sure, it has a high probability of being true, but maybe you are just experimenting with LSD and experiencing a vivid acid fantasy of God's perfect mathematical tome, or the CIA has placed a hidden video display in your eyeball replacing subversive *Math Mutation* literature with a fake, government-approved text designed to calm the masses. You cannot absolutely guarantee either of these hypotheses to be false, yet you carry on with your day based on the belief that you have in fact read the *Math Mutation* book. So where is the problem with our original premises?

Perhaps I'm not doing justice to the millions of words philosophers have written about this paradox, but coming from my computer science world, it looks to me like a simple case of a "rounding error". When we started off by saying we are justified in believing the lottery ticket would not win, we were really being lazy: the correct statement is "the ticket will lose with 99.9 % probability." By wanting to establish a binary yes-or-no belief, we are rounding that 99.9 % to 100 %, throwing away information. And as anyone who has dealt with calculations in computer systems can tell you, a rounding that is reasonable for a single data item can accumulate to a ridiculous error when many items are aggregated. Such accumulated round off errors in calculations have led to numerous real-life disasters, such as failures in the US Patriot missile system and in a Vancouver stock exchange index.

So this rounding to 100 % may be reasonable for a single ticket, but cause a catastrophic error when applied to 1000 tickets. Looked at in this way, the paradox is not a paradox at all, but just a consequence of the fact that calling something a 'belief' is a rounding of a more complex probabilistic quantity to a simple 0 % or 100 %. Sometimes we have to round a probability to a belief in order to act in real life – to make the decision whether or not to buy that lottery ticket, for example – but that's not a contradiction, just an approximation we use to conduct our daily business. Just about all scientific theories work this way: we are not 100 % sure any of them are true, but sure enough that we can tentatively act on the basis of their truth for now.

My favorite aspect of this lottery paradox, though, is the extension known as David Makinson's Preface Paradox. Suppose you have carefully researched every episode of, for example, a 200-episode podcast, and are pretty sure, but not 100 % sure, that each episode states the facts correctly as you produce it. You do know you're not perfect, and due to the number of episodes, chances are that one of them must have an error somewhere. If you have a 99 % accuracy rate, then you likely have released 2 wrong episodes somewhere along the line. So, when the bestselling book based on transcripts of your podcasts is being prepared for distribution, are you justified in writing in your preface that the book contains an error? While you rationally believe that each statement in the podcasts is correct, you also believe, based on the probabilities, that you made a mistake somewhere. Thus in a sense, you rationally believe that *Math Mutation* does and does not contain an error. Think about it for a few minutes, and I think you'll realize that the rounding explanation applies to this example as well: the best resolution is to say that, when you remove the rounding you did with each chapter, *Math Mutation* does probably contain an error somewhere. Sorry about that.

Does This Podcast Exist?
From Math Mutation podcast 177

Recently a friend pointed me to an interesting article at io9.com, titled "You're living in a computer simulation, and math proves it." I'm sure you're familiar with the concept of reality being an elaborate computer simulation, an idea most famously illustrated in the *Matrix* movies, but around in some form or another for many years before that. Even before computers, philosophers were discussing whether life was real or just some kind of dream in the mind of powerful unknown beings. A lot of the recent discussion of the topic seems to have been spurred by a 2003 Philosophical Quarterly article by Nick Bolstrom of Oxford University, "Are You Living in a Computer Simulation?", which was one of the inspirations for the io9 article. According to this particular article, we can prove through math alone that we are almost certainly living in a computer simulation.

Here's how the basic argument goes. Let's suppose it is possible for a sufficiently advanced civilization to create a computer capable of running a simulation of the complexity of our universe. If such a simulation could be created, would these advanced beings create only one, or many of them? Just look at the sales figure for *The Sims*, a much more primitive simulation game from our time, and I think you'll agree that if you could buy an *Xbox* disc that runs a full simulation of an Earth-like world for your own amusement, lots of people would do it. So for every real reality, there are millions or even billions of simulated realities. Thus, let's ask the question: given an arbitrary reality, namely ours, is it real or simulated? The odds are millions or billions to one that it's one of the simulated ones. Seems like a pretty convincing argument, doesn't it?

Of course, this is dependent on a few premises that may be a bit questionable. First is the idea that it would be possible to generate a simulation of the complexity of our reality: while extrapolating the rate of advances in computing over the last century makes it sound plausible, we can't really be sure. There may be some fundamental limit to computing power that falls short of what is needed for a full simulation of this type. Another premise is that creatures advanced enough to create such a simulation would choose to do it – maybe such advanced civilizations would tend to develop moral philosophies that wouldn't accept the creation of conscious beings for other's amusement. I'd be more inclined to argue from the reverse of this idea, actually: if some powerful beings could arbitrarily mess with the laws of our reality for their own amusement, like some gamer who uses *Fallout 3* cheat mods to turn everyone within a mile into two-headed cows just for fun, why aren't we observing a lot more arbitrary violations of physical laws?

A more serious hole in this argument, also discussed by Bolstrom, is that we haven't discussed how many universes exist where civilization never reaches the level capable of creating such a simulation. Maybe it's true that when a civilization reaches this level, it will create a billion simulations – but at the same time, perhaps only one in a billion billion civilizations reaches this level, which upends the whole argument. It could just be a natural tendency of sentient beings to kill each other off in nuclear wars way before they get to the level of developing a universal-simulation computer program. If this is the case, the odds are a billion to one that we are real rather than part of an elaborate simulation. Sorry, *Matrix* fans.

Another intriguing trend in this discussion is the search by some physicists for direct evidence that we are actually in a computer simulation. The idea is to find, in the actual physical laws of our universe, elements that would be telltale signs of a computer simulation. For example, scientists could detect effects of cosmic rays traveling in a tiny lattice of regular grid lines, like pixels on a computer screen, rather than being able to truly exist continuously in spacetime. It sounds like a nice idea, but I'm pretty skeptical of such a search: how can we say for sure that such lattice-like behavior, or any unexpected observation from physics, is the result of being a simulation rather than some subtle new law of physics we have not yet discovered? If they had observed relativistic effects experimentally before Einstein came along, would they claim that time distortion during high acceleration was proof that we were living in a buggy computer simulation, since the observations violated reality, which everyone knows to be described by Newton's laws?

Along similar lines are the arguments that we can infer proof of artificial creation of our universe by the fact that so many physical constants just happen to be fine-tuned to allow sentient life. This one comes up a lot in Creationism and "Intelligent Design" arguments as well. But as I see it, this line of reasoning has been thoroughly demolished by the anthropic principle: maybe an equal number of other universes with different constants do exist, but those of us asking the question have to be in the one with the life-friendly constants, otherwise we couldn't be asking it.

Anyway, I'm afraid this book will not be able to definitively answer the question of whether we are real or in a simulation. Also, if you're a meta-being listening to this podcast from another universe while monitoring our simulated universe on your computer screen, don't gloat too much at my amusing level of simulated ignorance: you might still be part of a simulation in a meta-meta-universe, and I can't definitively prove it either way for you either. But next time you're messing with our reality for fun, a few more 5-star podcast reviews on iTunes would be nice.

A Universe Made of Math
From Math Mutation podcast 105

You may have seen some discussion in the media of a bizarre theory by cosmologist Max Tegmark. He summarizes his theory with a concise description: "there is only mathematics; that is all that exists". In other words, the universe is only a mathematical object, and every consistent mathematical object is, in a sense, its own universe. If I understand him right, he is saying that the multiverse consists of all mathematically describable universes, and all have equal claim to existence.

If you aren't thoroughly confused yet, he goes on to specify that there are four 'levels' of the multiverse that we can describe. Level 1 is the infinite space that we seem to observe around us. Level 2 is the multiple set of Level 1 universes that arise if there is more than one solution to the Big Bang equations. Level 3 comes from the 'many worlds' interpretation of quantum mechanics, which we have discussed in previous chapters: whenever a quantum event occurs, where two possible outcomes can happen, the universe splits into multiple ones, each of which displays one of the outcomes. (This idea was originally proposed in 1957 by American physicist Hugh Everett, and further popularized by Bryce Seligman DeWitt in the 1960s, sadly no relation to me despite his cool name.) As Tegmark describes them, "The parallel universes are like different pages in a book, existing independently, simultaneously, and right next to each other. In a way all these infinite level III universes exist right here, right now."

Most bizarre is the Level 4 multiverse, which consists of all consistent mathematical structures which can describe a universe. In other words, any mathematically possible universe exists, in just as real a sense as our universe – we are just one of many possible sets of mathematical abstractions. Pretty crazy-sounding, but in a weird way, this does solve some core philosophical problems of physics. For one thing, when we discover various constants and equations that describe our universe, we always ask: why these? This hypothesis provides an answer: all possible sets of equations and constants that work form universes of their own, and we just happen to be observing the one we're in. And then the many-worlds interpretation of quantum mechanics, which calls for the universe to split into exponential numbers of others upon every quantum event, no longer causes a problem – instead of the impossible proposition of instantly duplicating physical universes, all the mathematically describable ones inherently come into existence.

I'm sure you have already come up with lots of critiques of this theory. We certainly feel like we exist; where does human consciousness fit in to all this? Can a mathematical abstraction really believe it is conscious? In a sense, though, if we think we can create artificial intelligence with computers, this isn't really that different. Another objection is: Is it really different to say that everything that is mathematically possible exists, and that nothing exists at all? This really reduces to the previous question in a sense: our own existence seems somehow qualitatively different than the 'existence' of mathematical objects we describe on paper.

The best objection I've heard to this theory overall is probably the "Principle of Finite Imagination": there is a logical reason why the universe is what it is and why we exist, though it may not be possible for us to understand. The fact that the best way we can describe our universe is through our system of mathematics may simply reflect the limitations of the human mind.

What do you think? Don't think too hard, because the answer you are bound to come up with may be pre-determined by the mathematical laws of the universe anyway.

Mathematical Immortality
From Math Mutation podcast 178

Ever since the dawn of mankind, and probably long before the first caveman figured out that 2 plus 2 equals 4, we have been wondering about what happens to our souls after we die. Does our consciousness continue in some form after we're gone? Or does it just disappear from the universe, never to be experienced again? I'm not going to try to answer this from the philosophical, spiritual or religious viewpoint, since there are thousands of better online flame wars available if we would like to debate those topics. But if we restrict ourselves to the domain of math and physics, can we still construct a good argument for some form of life after death?

One of the most direct arguments on this topic comes from the many worlds interpretation of quantum mechanics, as we discussed in the previous section. The many-worlds interpretation derives from the fact that quantum physics describes subatomic events in the form of probability waves, multiple possible outcomes with different probabilities. A definite outcome does not occur until the system is observed and the waveform collapses to a particular state. A classic example here is "Schrödinger's Cat", a cat locked in a box with a poison capsule set to release only if a particular radioactive atom decays in the next hour. (Luckily PETA was not yet around during the development of quantum physics.) Until the box is opened to collapse the wave function, the cat is neither alive nor dead: the box can only be described by a superposition of states that labels the cat possibly alive and possibly dead.

The many-worlds interpretation describes this situation in terms of multiple universes. When a quantum event occurs, such as the decay of the atom in the box, our universe actually forks into two universes, one representing each probability. So in one universe the cat is alive, and in the other the cat is dead. This particular example of the cat should make it clear how this relates to our discussion of immortality: while we may open the box, see the dead cat, and mourn it, there is another universe in which the cat jumped out of the box alive and is happily playing with a ball of USB cables. In any case where someone dies in a situation that they could have theoretically survived due to a differently-occurring quantum event, there is another universe nearby in which they are still alive.

Clever readers may have come up with an objection to claiming that this leads to a form of immortality: sure, it says something about life after premature deaths, but what about inevitable causes of death such as old age? Surely in every possible universe, as sad as it is for the podcasting community, I'll be dead 100 years from now. While previous generations are out of luck, we do have an answer for this one as well though: the technological singularity.

The singularity idea is that technology has been advancing at an exponential rate for most of the last century. Even when I was in college, I didn't dream that by middle age I would have a device in my pocket that could store a thousand science podcasts and still play *Walking Dead* games with computing power to spare. Technological growth has been in some ways like a graph of $y = 2^n$: while it may start slow, it very quickly starts verging towards infinity. If we can sustain this rate of growth, it may be only a matter of decades until we are capable of feats of technology that we would see as almost infinitely powerful by today's standards. When this "singularity" hits, we will be capable of uploading our brain's contents into a powerful computing device, continuing our consciousness indefinitely without the limitations of our frail human bodies. So if in just one parallel universe we can survive long enough to reach the singularity, and upload our brains into a computer with a good extended warranty and reliable backups, we truly are immortal.

If we're talking about the mind as software, we also should not discount another possible form of immortality, distinct from the many-universes-based ones we have been discussing until now. If your consciousness is formed by a finite pattern of electrical firings in your brain, why does this pattern have to occur only in your current brain and current body? There are things all over the universe: stars, planets, quasars, and other stuff, that constantly exhibit many arbitrary patterns of activity among the electrons that make them up. Why shouldn't one of these random patterns be effectively a software program that exactly executes your consciousness, except that it continues after your Earth-based body dies? Sure, this would be a bit of a coincidence – but if the universe is infinite, perhaps every possible electron pattern will occur somewhere and sometime. (There is the slight monkey wrench here of multiple different-sized infinities, as discussed earlier in this book, but since we don't know the proper classes of infinity for our brains' possibilities or the universe, we're free to speculate.) And this isn't even counting the possibility that we discussed a few sections back, where your mind is already an intentional simulation that can be rebooted.

Before we conclude, I'd like to dedicate this book to the memory of my father, Morton Seligman, who recently passed away at the age of 75. If he can just hold out until the singularity in a nearby universe, maybe we'll be seeing him again sometime. Or he might be sitting on an underworld throne next to Hades right now, laughing at us for these silly mathematical discussions while we ignore the true reality.

■ ■ ■

References and Further Reading

Here we provide a detailed list of the sources that inspired each chapter of this book, and that provided most of the original source data. Where a reference is specified as a web page with no clear date of origin, we date it for when the page was retrieved, during the editing of this book in early 2016. In a few unfortunate instances, some of my original podcast episode sources have disappeared from the web, but I have done my best to find substitute references in those cases.

Chapter 1: Simple Surprises

City of Mutants

No references.

Two Plus Two Equals Five

[And14] Craig Andresen, "Common Core is Nothing But Ideological Paste", The National Patriot, http://www.thenationalpatriot.com/tag/225/, 2014

[Ben13] J.R. Benjamin, "Does Two Plus Two Equal 4?", Bully Pulpit blog, http://jrbenjamin.com/2013/07/02/22/, 2013

[Des41] Rene Descartes, *Meditation 1*, as translated at http://www.wright.edu/~charles.taylor/descartes/meditation1.html. Originally published in 1641.

[Eul04] Houston Euler, "The History of 2+2=5", http://www.uni-graz.at/imawww/pages/humor/twoandtwo.html, 2004

[Pyu05] Eddie Pyune, "2+2=5", Urban Dictionary, http://www.urbandictionary.com/define.php?term=2+%2B+2+%3D+5, 2005

[Str98] Straight Dope, "Does 2+2=5 For Very Large Values of 2?", http://www.straightdope.com/columns/read/1382/does-2-2-5-for-very-large-values-of-2, 1998

[W2p16] Wikipedia article, "2+2=5", https://en.wikipedia.org/wiki/2_%2B_2_%3D_5,

Stupid Number Tricks

[Gar84] Martin Gardner, "Quiz Kids". New York Review of Books, March 16, 1984.
[Smi84] Stephen Smith, "Trade Secrets", New York Review of Books, November 8, 1984

Deceptive Digits

[Gol11] Ben Goldacre, "The Special Trick That Helps Us Identify Dodgy Stats", The Guardian, September 16 2011
[Hay12] Allyn H. Haynes, "Detecting Fraud in Bankrupt Municipalities Using Benford's Law", Scripps College thesis, http://scholarship.claremont.edu/cgi/viewcontent.cgi?article=1043&context=scripps_theses, 2012
[Nig99] Mark J. Nigrini, "I've Got Your Number", Journal of Accountancy, May 1999
[Tit12] John Titus, "How Obama is Guaranteeing the Next Financial Crisis will be Much Worse", The Daily Bail, http://dailybail.com/home/how-obama-is-guaranteeing-the-next-financial-crisis-will-be.html, 2012
[WBe16] Wikipedia article, "Benford's Law", http://en.wikipedia.org/wiki/Benford's_law

Nonrandom Randomness

[Ful08] John Fuller, "How the iPod Shuffle Works", How Stuff Works, http://electronics.howstuffworks.com/ipod-shuffle.htm, 2008
[WBi16] Wikipedia article, "Birthday Paradox", http://en.wikipedia.org/wiki/Birthday_paradox

A True Holiday Celebration

[Pau88] John Allen Paulos, Innumeracy, Hill and Wang, 1988
[Zyr16] Zyra, "Caesar's Last Breath", http://www.zyra.tv/lbreath.htm

Forgotten Knowledge

[Asi58] Asimov, Isaac, "The Feeling of Power", If: Worlds of Science Fiction, February 1958
[WEN16] Wikipedia article, "ENIAC", http://en.wikipedia.org/wiki/ENIAC

Exponents Squared

[Tet15] Tetration.org website, http://www.tetration.org/Tetration/index.html, 2015
[WTe16] Wikipedia article, "Tetration", http://en.wikipedia.org/wiki/Tetration

Giving You The Fingers

[Int02] Inutitor, "How to Count to 1023 on your Fingers", http://www.intuitor.com/counting/, 2002

[WCh16] Wikipedia article, "Chinese number gestures", http://en.wikipedia.org/wiki/Chinese_number_gestures

[Har16] Andy Harris, "Chisenbop Tutorial", http://www.cs.iupui.edu/~aharris/chis/chis.html

[Ias16] Iowa State University, "Medieval Finger Counting", http://orion.math.iastate.edu/mathnight/activities/modules/count/countleft.pdf

Chapter 2: Into Infinity

Too Infinite For Me

[WCa16] Wikipedia article, "Cantor's Diagonal Proof", http://en.wikipedia.org/wiki/Cantor%27s_diagonal_proof

[WHi16] Wikipedia article, "Hilbert Hotel", http://en.wikipedia.org/wiki/Hilbert_Hotel

Infinitely Ahead of His Time

[WGe16] Wikipedia article, "Georg Cantor", http://en.wikipedia.org/wiki/Georg_cantor

[WTr16] Wikipedia article, "Transfinite Number", http://en.wikipedia.org/wiki/Transfinite_number

[Hay09] Gary and Michael Picard Hayden, *This Book Does Not Exist,* Fall River Press, 2009

Infinite Infinities

[WPo16] Wikipedia article, "Power Set", http://en.wikipedia.org/wiki/Power_set

Infinity Times Infinity

[WSu16] Wikipedia article, "Surreal Number", http://en.wikipedia.org/wiki/Surreal_number

[MCC07] MarkCC, "Introducing The Surreal Numbers", Scientopia blog, http://scientopia.org/blogs/goodmath/2007/03/29/introducing-the-surreal-numbers-edited-rerun, March 29 2007

[Cho06] Paul Chou, "Surreal Numbers Presentation Outline", http://ocw.alfaisal.edu/NR/rdonlyres/Mathematics/18-304Spring-2006/D94A1BD0-FD02-41CD-9334-5CD0FE83D873/0/surreal_chou.pdf, May 2006

[Con01] John Conway, *On Numbers and Games*, A.K.Peters, 2000

[Pri16] Princeton University, John Conway faculty bio, http://www.princeton.edu/main/news/archive/S37/10/88C95/#comp000051b310d4000000025e0bd3

Infinite Perimeter, Finite Area

[WKo16] Wikipedia article, "Koch Snowflake", http://en.wikipedia.org/wiki/Koch_snowflake

[WFr16] Wikipedia article, "Fractal", http://en.wikipedia.org/wiki/Fractal

A Pretty Big Library

[Bor62] Jorge Luis Borges, "The Library of Babel", in *Labyrinths,* New Directions, 1962

[WLi16] Wikipedia artcile, "The Library of Babel", http://en.wikipedia.org/wiki/The_Library_of_Babel

Someone Knocked My 8 Over

[Ruk82] Rudy Rucker, *Infinity and the Mind*, Birkhauser, 1982.

[WIn16] Wikipedia article, "Infinity", http://en.wikipedia.org/wiki/Infinity
Rudy Rucker's infinity and the mind

Not Quite Infinity

[Bae13] John Baez, Blog Entry, https://plus.google.com/117663015413546257905/posts/KJTgfjkTZCQ, January 11 2013

[Gar77] Martin Gardner "Mathematical Games", Scientific American, Nov. 1977

[MOv12] Math Overflow, "Reconstructing the Argument that Yields Graham's Number", http://mathoverflow.net/questions/117006/reconstructing-the-argument-that-yields-grahams-number, 2012

[Wei16] Eric W. Weisstein,"Graham's Number," MathWorld--A Wolfram Web Resource, http://mathworld.wolfram.com/GrahamsNumber.html

[WGa16] Wikipedia article, "Graham's Number", http://en.wikipedia.org/wiki/Graham%27s_number

[WKn16] Wikipedia article, "Knuth's Up-Arrow Notation", http://en.wikipedia.org/wiki/Knuth%27s_up-arrow_notation

Big Numbers Upside Down

[Cra97] Richard Crandall, "The Challenge of Large Numbers", Scientific American, February 1997

[Mun16] Robert Munafo, "Notable Properties of Specific Numbers", http://mrob.com/pub/math/numbers-22.html

[Vil09] John Carl Villanueva, "How Many Atoms Are There in the Universe", Universe Today, http://www.universetoday.com/36302/atoms-in-the-universe/, 2009

[WJo16] Wikipedia article, "John Carter of Mars", http://en.wikipedia.org/wiki/John_Carter_of_Mars

[WUn16] Wikipedia article, "The Universe", http://en.wikipedia.org/wiki/Universe

Chapter 3: Getting Geometric

Something Euclid Missed

[WMo16] Wikipedia article, "Morley's Trisector Theorem", http://en.wikipedia.org/wiki/Morley%27s_trisector_theorem

[Mat16] Mathpages.com, "Morley's Trisector Theorem", http://www.mathpages.com/home/kmath376/kmath376.htm

[Bog16] Alexander Bogolmony, "Morley's Miracle", http://www.cut-the-knot.org/triangle/Morley/

[Ubc16] UBC Math Pages, "J. Conway's Proof", http://www.math.ubc.ca/~cass/courses/m308-02b/projects/hui/proof1.html

How Not To Decorate Your Bathroom

[WGi16] Wikipedia article, "Girih Tiles", http://en.wikipedia.org/wiki/Girih_tiles

[WPe16] Wikipedia article, "Penrose Tiling", http://en.wikipedia.org/wiki/Penrose_tiling

Bees Vs Mathematicians

[Pet99] Ivars Peterson, "The Honeycomb Conjecture", Science News, July 1999

[WHo16] Wikipedia article, "Honeycomb", http://en.wikipedia.org/wiki/Honeycomb

[WRe16] Wikipedia article, "Regular Tiling", http://en.wikipedia.org/wiki/Regular_tiling

A Brush With Evil

[WPe16] Wikipedia article, "Pentagram", http://en.wikipedia.org/wiki/Pentagram

[Wei16b] Eric W. Weisstein, "Pentagram," MathWorld--A Wolfram Web Resource. http://mathworld.wolfram.com/Pentagram.html

[Ske13] Skeptic's Dictionary, "The Pentagram", http://www.skepdic.com/pentagram.html

Twistier Than You Thought

[WMo16] Wikipedia article, "Mobius Strip", http://en.wikipedia.org/wiki/M%C3%B6bius_strip

[Ple16] David Pleacher, "Mobius Strip", Mr. P's Math Page, http://www.pleacher.com/mp/puzzles/tricks/mobistrp.html

[Doh00] Paul Doherty, "Mobius Dissection", http://www.exo.net/~pauld/activities/mobius/mobiusdissection.html, 2000

[Ger02] Vivienne Gerristen, "The Protein With a Topological Twist", Protein Spotlight, http://web.expasy.org/spotlight/back_issues/sptlt020.shtml, 2002

[WRe16] Wikipedia article, "Recycling Symbol", http://en.wikipedia.org/wiki/Recycling_symbol

[Thu16] Eric Thul, "The Geometry of Musical Chords", McGill University, http://www.cs.mcgill.ca/~ethul/pub/course/comp644/project/z-onepage.pdf

[Har09] Vi Hart, "Mobius Music Box", YouTube, http://www.youtube.com/watch?v=3iMI_uOM_fY, 2009

[Kas16] Alex Kasman, "Motif=Mobius Strip/Nonorientability", Mathematical Fiction, http://kasmana.people.cofc.edu/MATHFICT/search.php?orderby=title&go=yes&motif=mob

Squash Those Dice

[WPl16] Wikipedia article, "Platonic Solids", http://en.wikipedia.org/wiki/Platonic_solids

Crazy Dice

[Dis16] The Dice Shop, "D7 Specialist Dice", https://www.thediceshoponline.com/dice-sets/157/D7-Specialist-Dice

[Dic16] Dice Collector, "Dice Patents", http://www.dicecollector.com/DICEINFO_PATENTS.html

Wheels That Aren't Round

[San08] Chris Sangwin, "Applications of Shapes of Constant Width", http://web.mat.bham.ac.uk/C.J.Sangwin/howroundcom/roundness/applications.html, 2008

[Gar14] Martin Gardner, *Knots and Borromean Rings,* Cambridge University Press, 2014

[WCu16] Wikipedia article, "Curve of Constant Width", https://en.wikipedia.org/wiki/Curve_of_constant_width

[WRe16] Wikipedia article, "Reuleaux Triangle", https://en.wikipedia.org/wiki/Reuleaux_triangle

The Future That Never Arrived

[WGe16] Wikipedia article, "Geodesic Dome", http://en.wikipedia.org/wiki/Geodesic_dome

[Bal16] J. Baldwin, "Geodesic Domes", http://www.thirteen.org/bucky/dome.html

[BFI16] The Buckminster Fuller Institute, http://bfi.org

[Fie16] Simon Field, "A Geodesic Dome", Sci Toys, http://sci-toys.com/scitoys/scitoys/mathematics/dome/dome.html

[WBu16] Wikipedia article, "Buckminster Fuller", http://en.wikipedia.org/wiki/Buckminster_Fuller

[Urn99] Kirby Urner, "Synergetics on the Web", http://www.grunch.net/synergetics/, 1999

Chapter 4: Deeper Dimensions
Making Flatland Real

[Abb80] Edwin A. Abbott, *Flatland,* 1884, full text available at Gutenberg: http://www.gutenberg.org/ebooks/97

[Bur83] Dionys Burger, *Sphereland,* Harper and Row, 1983

[Dew00] A.K. Dewdney, *The Planiverse,* Copernicus, 2000

[Hin80] C.H. Hinton, *Speculations on the Fourth Dimension,* Dover, 1980

[Ste08] Ian Stewart, *The Annotated Flatland*, Basic Books, 2008

[WFl16] Wikipedia article, "Flatland", http://en.wikipedia.org/wiki/Flatland

Is Flatland Doomed?

[Dew00] A.K. Dewdney, *The Planiverse,* Copernicus, 2000

[WEs16] Wikipedia article, "Escape Velocity", http://en.wikipedia.org/wiki/Escape_velocity

Visitors From The Next Dimension

[Ruc85] Rudy Rucker, *The Fourth Dimension*, Mariner Books, 1985

[WFo16] Wikipedia article, "The Fourth Dimension", http://en.wikipedia.org/wiki/Fourth_dimension

Will The Real Fourth Dimension Please Stand Up?

[Ruc85] Rudy Rucker, *The Fourth Dimension*, Mariner Books, 1985

[WDi16] Wikipedia article, "Dimension", http://en.wikipedia.org/wiki/Dimension

[WWo16] Wikipedia article, "Worldlines", http://en.wikipedia.org/wiki/Worldline

A Four Dimensional House

[Hei41] Robert Heinlein, "And He Built a Crooked House", Astounding Science Fiction, February 1941

Turning Around In Time

[Ruc85] Rudy Rucker, *The Fourth Dimension*, Mariner Books, 1985

[Har99] David M. Harrison, "AntiMatter", Upscale, http://www.upscale.utoronto.ca/GeneralInterest/Harrison/AntiMatter/AntiMatter.html, 1999

[New97] New Scientist, "World of Antimatter", https://www.newscientist.com/article/mg15320699.400-world-of-antimatter/, 1997

11-Dimensional Spaghetti Monsters

[WSt16] Wikipedia article, "String Theory", http://en.wikipedia.org/wiki/String_theory

Your Five Dimensional Kitchen

[Cyb16] Cybernox Cookware site, http://www.dvorsons.com/Sitram/Cybernox/Cybernox.html

[Sli06] Slidefinder, "Quasicrystals", http://www.slidefinder.net/q/quasicrystals_what_why_exist/northwestern_4-7-2006/7508252, 2006

[WPe16] Wikipedia article, "Penrose Tiling", http://en.wikipedia.org/wiki/Penrose_tiling

[WQu16] Wikipedia article, "Quasicrystals", http://en.wikipedia.org/wiki/Quasicrystal

[Web16] Steffan Weber, "Quasicrystals web page", http://jcrystal.com/steffenweber/qc.html

As Math Goes By

[Dav08] Philip J. Davis, "Will A Kiss Remain A Kiss?", SIAM news, Vol 41, No 8, April 2008

[Ran06] Lisa Randall, *Warped Passages*, Harper Perennial, 2006

[Hup31] Herman Hupfeld, "As Time Goes By", available at http://www.reelclassics.com/Movies/Casablanca/astimegoesby-lyrics.htm, 1931

[Ste16] Rod Stewart, "As Time Goes By" alternate lyrics, http://www.azlyrics.com/lyrics/rodstewart/astimegoesby.html

Between The Dimensions

[Van16] Vanderbilt University, "Fractals and the Fractal Dimension", http://www.vanderbilt.edu/AnS/psychology/cogsci/chaos/workshop/Fractals.html

Chapter 5: Understanding The Universe

The Bogus Bang?

[Ruc08] Rudy Rucker, "Voices in the White", Rudy Rucker Blog, http://www.rudyrucker.com/blog/2008/05/05/voices-in-the-white/, May 5th 2008

[Ste16] Paul J. Steinhardt, "A Brief Introduction to the Ekpyrotic Universe", Princeton University, http://www.physics.princeton.edu/~steinh/npr/

[Whi01] David Whitehouse, "Before the Big Bang", BBC News, http://news.bbc.co.uk/2/hi/science/nature/1270726.stm, 2001

The Shape of the Universe

[Bat04] Stephen Battersby, "Big Bang Glow Hints at Funnel-Shaped Universe", New Scientist, https://www.newscientist.com/article/dn4879-big-bang-glow-hints-at-funnel-shaped-universe/, 2004

[Lum06] Jean-Pierre Luminet, "The Shape of the Universe After WMAP Data", Brazilian Journal of Physics, Vol 36, No 1b, March 2006

[Nas14] NASA, "Will The Universe Expand Forever?", http://map.gsfc.nasa.gov/universe/uni_shape.html, 2014

[WPi16] Wikipedia article, "Picard Horn", https://en.wikipedia.org/wiki/Picard_horn

[WSh16] Wikipedia article, "Closed Universe", http://en.wikipedia.org/wiki/Closed_universe#Closed_Universe

Your Size in Space and Time

[Gru88] Robert Grudin, *Time and the Art of Living*, Ticknor & Fields, 1988

Observing the Universe

[WOb16] Wikipedia article, "Observable Universe", http://en.wikipedia.org/wiki/Observable_universe

Alien Algebra

[Pom05] Carl Pomerance, "Prime Numbers and the Search for Extraterrestrial Intelligence", https://math.dartmouth.edu/~carlp/PDF/extraterrestrial.pdf, 2005

Time Reversed Worlds

[Gar06] Martin Gardner, *The New Ambidextrous Universe,* Dover, 2006

[Quo12] Quora, "Why is it believed that positrons are electrons that are moving backwards in time?", https://www.quora.com/Why-is-it-believed-that-positrons-are-electrons-that-are-moving-backwards-in-time, 2012

[Von69] Kurt Vonnegut, *Slaughterhouse-Five*, Delacorte, 1969

[WAn16] Wikipedia article, "Antimatter", http://en.wikipedia.org/wiki/Antimatter

[WBa16] Wikipedia article, "Baryon Asymmetry", http://en.wikipedia.org/wiki/Baryon_asymmetry

[WFe16] Wikipedia article, "Feynman Diagram", http://en.wikipedia.org/wiki/Feynman_diagram

A Pear Shaped Planet

[Roi03] Ron Roizen, "Christopher Columbus, Magnificent Bungler", Shoshone News-Press, http://www.roizen.com/ron/Columbus.htm, October 12 2003

[Sac16] Sacred Texts, "The Earth of Columbus", http://www.sacred-texts.com/earth/boe/boe26.htm

[WFl16] Wikipedia article, "Myth of the Flat Earth", http://en.wikipedia.org/wiki/Myth_of_the_Flat_Earth

[WGe16] Wikipedia article, "Geodesy", http://en.wikipedia.org/wiki/History_of_geodesy

[WMa16] Wikipedia article, "Marinus of Tyre", http://en.wikipedia.org/wiki/Marinus_of_Tyre

Where Am I?

[Fer02] J. Donald Fernie, "Finding Out The Longitude", American Scientist, http://www.americanscientist.org/issues/pub/2002/9/finding-out-the-longitude/2, 2002

[WHi16] Wikipedia article, "History of Longitude", http://en.wikipedia.org/wiki/History_of_longitude

[WJo16] Wikipedia article, "John Harrison", http://en.wikipedia.org/wiki/John_Harrison

Putting the Multiverse to Work

[Deu97] David Deutsch, *The Fabric of Reality*, Viking Adult, 1997

[Gar03] Martin Gardner, *Are Universes Thicker Than Blackberries*, W. W. Norton and Company, 2003

[WMa] Wikipedia article, "Many Worlds Interpretation", http://en.wikipedia.org/wiki/Many-worlds_interpretation

[WQu] Wikipedia article, "Quantum Computing", http://en.wikipedia.org/wiki/Quantum_computing

Chapter 6: The Mathematical Mind
What Color Is This Podcast?

[Sci06] Scientific American, "What Is Synesthesia?", http://www.scientificamerican.com/article/what-is-synesthesia/

[WNa16] Wikipedia article, "Vladimir Nabokov", http://en.wikipedia.org/wiki/Vladimir_Nabokov

[WSy16] Wikipedia article, "Synesthesia", http://en.wikipedia.org/wiki/Synesthesia

Computers On The Brain

[Max06] Clive Maxfield, "A Reader Responds to 'Seeing Schematics in Color'", EE Times, http://www.eetimes.com/author.asp?section_id=14&doc_id=1282735, September 21 2006

[Max07] Clive Maxfield, "Max's Chips and Dips: Seeing Black-And-White Schematics In Color", Chip Design Magazine, http://chipdesignmag.com/print.php?articleId=1518?issueId=24, August/September 2007

The Rain Man's Secret

[Tam07] Daniel Tammet, *Born on a Blue Day*, Free Press, 2007

Look Him In The Eye

[Rob08] John Elder Robison, *Look Me In The Eye*, Three Rivers Press, 2008

[Rob16] John Elder Robison's Website, http://www.johnrobison.com

Savants Are People Too

[Tam09] Daniel Tammet, *Embracing the Wide Sky*, Atria Books, 2009

[Tam16] Daniel Tammet, "Optimnem Website", http://www.optimnem.co.uk

The Uninhibited Brain

[Tam09] Daniel Tammet, *Embracing the Wide Sky*, Atria Books, 2009

[Tam16] Daniel Tammet, "Optimnem Website", http://www.optimnem.co.uk

A Logical Language

[Log16] Loglan official website, http://www.loglan.org

[Loj16] Lojban official website, http://lojban.org

[Nic16] Nick Nicholas and John Cowan, "What Is Lojban?", http://jbotcan.org/whatislojban/book1.html

[Nvg16] NVG, "Why I Like Lojban", NVG Lojban page, http://arj.nvg.org/lojban/why-i-like.html

[Nvg16b] NVG, "Reasons Not To Like Lojban", NVG Lojban page, http://arj.nvg.org/lojban/anti-lojban.html

[Rat16] Rational wiki, "Lojban", http://rationalwiki.org/wiki/Lojban

[WEs16] Wikipedia article, "Esperanto", http://en.wikipedia.org/wiki/Esperanto

[WKl16] Wikipedia article, "Klingon Language", http://en.wikipedia.org/wiki/Klingon_language

[WLo16] Wikipedia article, "Loglan", http://en.wikipedia.org/wiki/Loglan

When 'Is' Isn't

[Bou89] D. David Bourland, "To Be or Not To Be: E-Prime as a Tool for Critical Thinking", ETC: A Review of General Semantics, Vol. 46, No. 3, Fall 1989

[Fre11] James D. French, "The Top Ten Arguments Against E Prime", General Semantics website, http://www.generalsemantics.org/wp-content/uploads/2011/05/articles/etc/49-2-french.pdf, 2011

[Her16] John C. Herbert, "English Prime as an Instructional Tool in Writing Classes", http://web.archive.org/web/20061007112531/http://exchanges.state.gov/forum/vols/vol41/no3/p26.htm

[WEp16] Wikipedia article, "E Prime", http://en.wikipedia.org/wiki/E-Prime

[Zim03] Daniel Zimmerman, "E Prime as a Revision Strategy", http://www.ctlow.ca/E-Prime/zimmerman.html, 2003

De-Abstracting Your Life

[AGS16] Australian General Semantics Society website, http://members.pcug.org.au/~ajames/agsBenefits.htm

[Gar57] Martin Gardner, *Fads and Fallacies in the Name of Science*, Dover, 1957

[IGS16] Institute of General Semantics website, http://www.generalsemantics.org

[War12] Ken Ward, "General Semantics", http://www.trans4mind.com/personal_development/KenGenSemantics.htm, 2012

[WGe16] Wikipedia article, "General Semantics", https://en.wikipedia.org/wiki/General_semantics

Your Kids Are Smarter Than You

[Chi14] Tom Chivers, "The Flynn Effect: Are We Really Getting Smarter?", Telegraph, http://www.telegraph.co.uk/news/science/science-news/11200900/The-Flynn-effect-are-we-really-getting-smarter.html, October 31 2014

[Gam12] Megan Gambino, "Are You Smarter Than Your Grandfather?", Smithsonian, http://www.smithsonianmag.com/science-nature/are-you-smarter-than-your-grandfather-probably-not-150402883, December 3 2012

[Gla07] Malcolm Gladwell, "None of the Above", New Yorker, December 17 2007

[Gou96] Stephen Jay Gould, *The Mismeasure of Man*, W. W. Norton & Company, 1996

[Tra14] Lisa Trahan, Karla K. Stuebing, Merril K. Hiscock, and Jack M. Fletcher, "The Flynn Effect: A Meta-Analysis", National Institutes of Health, http://www.ncbi.nlm.nih.gov/pmc/articles/PMC4152423/, 2014

[WFl16] Wikipedia article, "Flynn Effect", http://en.wikipedia.org/wiki/Flynn_effect

My Brain Hurts

[Hof01] Douglas R. Hofstadter and Daniel C. Dennett, *The Mind's I,* Basic Books, 2001

Psychochronometry

[Ken02] James Main Kenney, "Logtime: The Subjective Scale of Life", http://www.kafalas.com/Logtime.html, 2002

[Tam14] Daniel Tammet, *Thinking In Numbers,* Back Bay Books, 2014

[Tay07] Steve Taylor, "The Speed of Life", New Dawn Magazine, http://www.newdawnmagazine.com/articles/the-speed-of-life-why-time-seems-to-speed-up-and-how-to-slow-it-down, September 1 2007

Chapter 7: Science and Skepticism
Why Statisticians Stink at Statistics

[Tal10] Nassim Nicholas Taleb, *The Black Swan*, Random House, 2010

[WCo16] Wikipedia article, "Cognitive Bias", http://en.wikipedia.org/wiki/Cognitive_bias

On Average, Things Are Average

[WRe16] Wikipedia article, "Regression Towards the Mean", http://en.wikipedia.org/wiki/Regression_towards_the_mean

Don't Panic

[Bla09] Michael Blastland and Andrew Dilnot, "The Last Word: When Numbers Deceive", The Week, http://www.theweek.com/articles/508368/last-word-when-numbers-deceive, 2009

It Must Be True, There's An Equation

[Sok99] Alan Sokal and Jean Bricmont, *Fashionable Nonsense*, Picador, 1999

[Tri16] Tripod.com, "Euler's Proof of God", http://leonhard-euler.tripod.com/id4.html

[WAl16] Wikipedia article, "Alan Sokal", https://en.wikipedia.org/wiki/Alan_Sokal

A Twisted Take on Turing

[SCI16] SCIgen official site, https://pdos.csail.mit.edu/archive/scigen/

[WCo16] Wikipedia article, "Context-Free Grammar", https://en.wikipedia.org/wiki/Context-free_grammar

[WSc16] Wikipedia article, "SCIgen", https://en.wikipedia.org/wiki/SCIgen

I Want My Molecule

[WAv16] Wikipedia article, "Avogadro's Number", http://en.wikipedia.org/wiki/Avogadro%27s_number

[WHo16] Wikipedia article, "Homeopathy", http://en.wikipedia.org/wiki/Homeopathy

The Gullible Ratio

[Ama16] Amazon.com book page, "The Golden Ratio Lifestyle Diet", http://www.amazon.com/The-Golden-Ratio-Lifestyle-Diet-ebook/dp/B00FO810VQ

[Gar96] Martin Gardner, *Weird Water and Fuzzy Logic,* Prometheus, 1996

[See16] Nick Seewald, "The Myth of the Golden Ratio", http://goldenratiomyth.weebly.com/phi-in-psychology.html

[Spi16] Mike Spinak, "The Golden Section Hypothesis: A Critical Look", Naturography, http://naturography.com/the-golden-section-hypothesis-a-critical-look/

[WGo16] Wikipedia article, "Golden Ratio", http://en.wikipedia.org/wiki/Golden_ratio

Monkeying Around With Probability

[Kim16] John Kimball, "The Origin of Life", http://users.rcn.com/jkimball.ma.ultranet/BiologyPages/A/AbioticSynthesis.html

[Olo08] Peter Oloffson, "Probability, Statistics, Evolution, and Intelligent Design", Talk Reason, http://www.talkreason.org/articles/chanceprob.cfm, November 24 2008

[Pau06] John Allen Paulos, "What's Wrong With Creationist Probability?", ABC News, http://abcnews.go.com/Technology/story?id=2384584&page=1, September 3 2006

[Wat02] Brett Watson, "A Tiny Mathematical Proof Against Evolution", Free Republic, http://www.freerepublic.com/focus/news/640506/posts, 2002

Solving Burma's Problems

[Gur03] Mohan Guruswamy, "A Review of India's Policy on Burma", Mizzima News, http://burmatoday.net/mizzima2003/mizzima/2003/06/030613_india_mizzima.htm, June 13 2003

[WNi16] Wikipedia article, "9", https://en.wikipedia.org/wiki/9_(number)

A New Numerology

[WHu16] Wikipedia article, "Human Genome", http://en.wikipedia.org/wiki/Human_genome

[WNu16] Wikipedia article, "Numerology", http://en.wikipedia.org/wiki/Numerology

One Intestinal Worm Per Child

[Ogd09] Timothy Ogden, "Computer Error?", Pacific Standard, http://www.psmag.com/business-economics/computer-error-3520, August 20 2009

[WAm16] Wikipedia article, "Amdahl's Law", https://en.wikipedia.org/wiki/Amdahl%27s_law

[WOn16] Wikipedia article, "One Laptop Per Child", http://en.wikipedia.org/wiki/One_laptop_per_child

Chapter 8: Analyzing The Arts
Discovering the Third Dimension

[Aut12] Automotive Illustrations, "Perspective Drawing Basics", http://www.automotiveillustrations.com/tutorials/perspective-drawing-basics.html

[Sup01] Patrick Suppes, "Finitism in Geometry", http://suppes-corpus.stanford.edu/articles/ll/386.pdf

[WAg16] Wikipedia article, "Agatharcus", http://en.wikipedia.org/wiki/Agatharchus

[WEb16] Web Exhibits, Pompeii Image, http://www.webexhibits.org/sciartperspective/i/perspective1_large-01.jpg

[WPe16] Wikipedia article, "Perspective (graphical)", http://en.wikipedia.org/wiki/Perspective_(graphical)

A New Perspective on Perspective

[Gun08] Ben Gunnink, "Technology as Symptom and Dream", http://dreamflesh.com/library/robert-d-romanyshyn/technology-as-symptom-and-dream/, 2008

[WMa16] Wikipedia article, "Marshall McLuhan", http://en.wikipedia.org/wiki/Marshall_McLuhan

Hippasus's Revenge

[Boy16] John Boyd-Brent, "Pythagoras: Music and Space", About Scotland, http://www.aboutscotland.co.uk/harmony/prop.html

[Nov16] Yuval Nov, "Explaining the Equal Temperament", http://www.yuvalnov.org/temperament/

[WEq16] Wikipedia article, "Twelve Tone Equal Temperament", http://en.wikipedia.org/wiki/Twelve_tone_equal_temperament

Unlistenable But Fun

[Cag04] John Cage, *John Cage: Writer*, Limelight, 2004

[WJo16] Wikipedia article, "John Cage", http://en.wikipedia.org/wiki/John_Cage

What a Planet Sounds Like

[Lou10] Louise, "Carmen of the Spheres", Music and Astronomy, http://musicandastronomy.blogspot.com/2010/02/carmen-of-spheres.html, February 4 2010

[WMu16] Wikipedia article, "Music of the Spheres", http://en.wikipedia.org/wiki/Music_of_the_spheres

[WOc16] Wikipedia article, "Octave", http://en.wikipedia.org/wiki/Octave

Mozart Rolls the Dice

[Chu95] John Chuang, "Mozart's Musikalisches Würfelspiel", http://sunsite.univie.ac.at/Mozart/dice/, 1995

[Wag00] Willis Wager, "Composers and Crapshooters", Carousel Music, http://www.carousel-music.com/shooters.html, 2000

[WMo16] Wikipedia article, "Musikalisches Würfelspiel", http://en.wikipedia.org/wiki/Musikalisches_Würfelspiel

Candide's Calculus

[Vol59] Voltaire, *Candide*, originally published in 1759, full English translation available from Project Gutenberg at http://www.gutenberg.org/files/19942/19942-h/19942-h.htm

[WCa16] Wikipedia article, "Candide", http://en.wikipedia.org/wiki/Candide

[WGo16] Wikipedia article, "Gottfried Leibniz", http://en.wikipedia.org/wiki/Gottfried_Leibniz

Fractals in the Hat

[Lak93] Ahklesh Lakhtakia, "Fractals and The Cat in the Hat", Penn State News, http://news.psu.edu/story/140794/1993/09/01/research/fractals-and-cat-hat, September 1 1993

[Nel07] Philip Nel, *The Annotated Cat*, Random House, 2007

[Seu58] Dr Seuss, *The Cat in the Hat Comes Back*, Random House, 1958

[WFr16] Wikipedia article, "Fractals", http://en.wikipedia.org/wiki/Fractals

Math or Not Math?

[Hof98] Douglas R. Hofstadter, *Le Ton Beau de Marot*, Basic Books, 1998

Gnarly Gnovels

[Ruc13] Rudy Rucker, "What is Gnarl?", Rudy Rucker Blog, http://www.rudyrucker. com/blog/2013/09/06/gnarly-sf-reality-1-what-is-gnarl/, September 6 2013

[Von69] Kurt Vonnegut, *Slaughterhouse-Five*, Delacorte, 1969

[WKo16] Wikipedia article, "Kolmogorov Complexity", http://en.wikipedia.org/ wiki/Kolmogorov_complexity

Chapter 9: Political Ponderings

A Founding Theorem

[WMa16] Wikipedia article, "Marquis de Condorcet", http://en.wikipedia.org/ wiki/Marquis_de_Condorcet

More Than a Cartoon Cat

[Kle95] Suzanne Klebe, "President Garfield and the Pythagorean Theorem", American Almanac, http://american_almanac.tripod.com/garfield.htm, February 1995

[WJa16] Wikipedia article, "James A. Garfield", http://en.wikipedia.org/wiki/ James_A._Garfield

The Round Road to Damnation

[Ada91] Cecil Adams, "Did a state legislature once pass a law saying pi equals 3?", Straight Dope, http://www.straightdope.com/columns/read/805/did-a-state-legislature-once-pass-a-law-saying-pi-equals-3, February 22 1991

[WPi16] Wikipedia article, "Pi", http://en.wikipedia.org/wiki/Pi

A Math Teacher to Remember

[Phi08] Tony Phillips, "Solzhenitsyn Mathematician", American Mathematical Society, http://www.ams.org/news/math-in-the-media/mmarc-10-2008-media, October 2008

[WSo16] Wikipedia article, "Solzhenitsyn", http://en.wikipedia.org/wiki/Solzhenitsyn

Election Solutions

[McC12] Amy McCaig, "Hand Counts of Votes May Cause Errors, Says New Rice Study", Rice University, http://news.rice.edu/2012/02/02/hand-counts-of-votes-may-cause-errors-says-new-rice-u-study/, February 2 2012

[Sei11] Charles Seife, *Proofiness*, Penguin Books, 2011

Democracy Doesn't Work

[Bou13] Don Boudreaux, "The Collective Is Not a Relevant Alternative to the Individual", Cafe Hayek, http://cafehayek.com/2013/11/the-collective-is-not-a-relevant-alternative-to-the-individual.html, November 7 2013

[Tal08] Presh Talwakar, "Arrow's Impossibility Theorem and the Voting Paradox", Mind Your Decisions, http://mindyourdecisions.com/blog/2008/02/12/game-theory-tuesdays-someone-is-going-to-be-unhappy-an-illustration-of-the-voting-paradox/, February 12 2008

[WAr16] Wikipedia article, "Arrow's Impossibility Theorem", http://en.wikipedia.org/wiki/Arrow%27s_impossibility_theorem

Drop That Number or We'll Shoot

[Loh06] Fred von Lohmann and Wendy Seltzer, "Death by DMCA", IEEE Spectrum, http://spectrum.ieee.org/computing/software/death-by-dmca, June 1 2006

[Tou08] David Touretzky, "Gallery of CSS Descramblers", http://www.cs.cmu.edu/~dst/DeCSS/Gallery/

[WDM16] Wikipedia article, "DMCA", http://en.wikipedia.org/wiki/DMCA

That's How We Do It In Government

[WMa16] Wikipedia article, "Margin of Error", http://en.wikipedia.org/wiki/Margin_of_error

Chapter 10: Money Matters
Tally Folly

[Bir16] Dave Birch, "Tallies & Technologies", http://www.arraydev.com/commerce/JIBC/9811-11.htm

[WTa16] Wikipedia article, "Tally Stick", http://en.wikipedia.org/wiki/Tally_stick

How to Bankrupt Your Boss and Get Rich

[Tal05] Nassim Nicholas Taleb, *Fooled By Randomness*, Random House, 2005
[Tal16] Nassim Nicholas Taleb's official website, "Fooled By Randomness", http://www.fooledbyrandomness.com

What Color Is Your Swan?

[Tal10] Nassim Nicholas Taleb, *The Black Swan*, Random House, 2010

Comparative Disadvantage

[Lan07] Lauren F. Landsburg, "Comparative Advantage", Library of Economics and Liberty, http://www.econlib.org/library/Topics/Details/comparativeadvantage.html, 2007
[Tal10] Nassim Nicholas Taleb, *The Black Swan*, Random House, 2010
[WCo16] Wikipedia article, "Comparative Advantage", http://en.wikipedia.org/wiki/Comparative_advantage
[WIr16] Wikipedia article, "Irish Potato Famine", http://en.wikipedia.org/wiki/Irish_potato_famine

Money for Math

[Cla16] The Clay Mathematics Institute official website, http://www.claymath.org
[Kla13] Erica Klarreich, "Unheralded Mathematician Bridges the Prime Gap", Quanta, https://www.quantamagazine.org/20130519-unheralded-mathematician-bridges-the-prime-gap/, May 19 2013
[Sin16] Simon Singh, "The Wolfskehl Prize", http://simonsingh.net/media/articles/maths-and-science/the-wolfskehl-prize/
[WFe16] Wikipedia article, "Fermat's Last Theorem", http://en.wikipedia.org/wiki/Fermat's_Last_Theorem

Liking the Lottery

[Bay14] Bay News, "10 Things More Likely to Happen Than Winning the Mega Millions Lottery", http://www.baynews9.com/content/news/baynews9/news/article.html/content/news/articles/cfn/2013/12/17/mega_millions_odds.html?cid=rss, December 17 2013
[Han12] Andrew Hanson, "Powerball: Why the Lottery Is Idiotic", http://mic.com/articles/7169/powerball-why-the-lottery-is-idiotic#.YrOMSiyJX, November 25 2012
[Hic13] Walter Hickey, "According to Math, Here's When You Should Buy a Powerball Ticket", Business Insider, http://www.businessinsider.com/heres-when-math-says-you-should-start-to-care-about-powerball-2013-9, September 16 2013
[New92] New York Times, "Group Invests $5 Million to Hedge Bets in Lottery", February 25 1992

[Sal07] Alan Salzberg, "The Lottery: A Tax on Stupidity?", http://what-are-the-chances.blogspot.com/2007/12/lottery-tax-on-stupidity.html, December 13 2007

[WEx16] Wikipedia article, "Lottery: Expected Value", http://en.wikipedia.org/wiki/Lottery#Expected_value

Number Nonsense

[Aye16] Nancy Ayers, "Calculators in the Classroom", http://www.math.wichita.edu/history/topics/calculators.html#calc

[Mat96] Mathematically Correct, "A Horizon Without Calculators", http://www.mathematicallycorrect.com/horizon.htm

Does The House Always Win?

[Mez03] Ben Mezrich, *Bringing Down the House*, Atria Books, 2003

Mutating Up the Corporate Ladder

[WFr16] Wikipedia article, "Fractals", http://en.wikipedia.org/wiki/Fractals

[Whe06] Margaret Wheatley, *Leadership and the New Science*, Berrett-Koehler Publishers, 2006

[Whe16] Margaret Wheatley's official website, http://margaretwheatley.com

The Converse of a CEO

[WIn16] Wikipedia article, "In Search of Excellence", http://en.wikipedia.org/wiki/In_Search_of_Excellence

Chapter 11: Looking at Life
Florence Nightingale, Math Geek

[Lip93] Sally Lipsey, "Mathematical Education in the Life of Florence Nightingale", https://www.agnesscott.edu/Lriddle/WOMEN/night_educ.htm, 1993

[Neu03] D. Neuhauser, "Florence Nightingale Gets No Respect", National Institutes of Health, http://www.ncbi.nlm.nih.gov/pmc/articles/PMC1743730/pdf/v012p00317.pdf, 2003

[WFl16] Wikipedia article, "Florence Nightingale", http://en.wikipedia.org/wiki/Florence_Nightingale

The Genius Who Cheated

[Gel12] Andrew Gelman, "Gregor Mendel's Suspicious Data", http://andrewgelman.com/2012/08/08/gregor-mendels-suspicious-data/, August 8 2012

[Pil84] Ira Pilgrim, "The Too-Good-To-Be-True Paradox and Gregor Mendel", The Journal of Heredity, http://irapilgrim.mcn.org/men02.html, 1984

[WGr16] Wikipedia article, "Gregor Mendel", http://en.wikipedia.org/wiki/Gregor_mendel

Shuttle Butt

[Ada00] Cecil Adams, "Was Standard Railroad Gauge Determined by Roman Chariot Ruts?", Straight Dope, http://www.straightdope.com/columns/read/2538/was-standard-railroad-gauge-48-determined-by-roman-chariot-ruts, February 18 2000

[Sno14] Snopes, "Horse's Pass", http://www.snopes.com/history/american/gauge.asp, January 14 2014

Booms and Busts

[Bra10] Larry Bradley, "Logistic Equation", http://www.stsci.edu/~lbradley/seminar/logdiffeqn.html

[Dal16] Richard Dallaway, "Chaos in the Pond", http://www.dallaway.com/pondlife/

[Mis16] Missouri Department of Conservation, "Better Rabbit Habitat", http://mdc.mo.gov/your-property/wildlife-your-property/small-game-your-property/better-rabbit-habitat

[WLo16] Wikipedia article, "Logistic Function", https://en.wikipedia.org/wiki/Logistic_function

The Boids and Bees of Leadership

[Fis09] Len Fisher, *The Perfect Swarm*, Basic Books, 2009

[Rey07] Craig Reynolds, "Boids", http://www.red3d.com/cwr/boids/

[Sch08] Kevin M. Schultz, Kevin M. Passino, and Thomas D. Seeley, "The mechanism of flight guidance in honeybee swarms: subtle guides or streaker bees?", Journal of Experimental Biology, http://jeb.biologists.org/content/jexbio/211/20/3287.full.pdf, 2008

[WBo16] Wikipedia article, "Boids", http://en.wikipedia.org/wiki/Boids

A New Kind of Decade

[WCo16] Wikipedia article, "Conway's Game of Life", http://en.wikipedia.org/wiki/Conway%27s_Game_of_Life

[Wol02] Stephen Wolfram, *A New Kind of Science*, Wolfram Media, 2002

[Wol12] Stephen Wolfram, "It's Been Ten Years", http://blog.stephenwolfram.com/2012/05/its-been-10-years-whats-happened-with-a-new-kind-of-science/, May 2012

Basic Bugs

[Bra86] Valentino Braitenburg, *Vehicles: Experiments in Synthetic Psychology*, Bradford Books, 1986

[WBr16] Wikipedia article, "Braitenburg Vehicle", http://en.wikipedia.org/wiki/Braitenberg_vehicle

[Wis99] John Wiseman, "Braitenburg Vehicles", http://people.cs.uchicago.edu/~wiseman/vehicles/, May 6 1999

Bugged By Math

[Mur16] Murderous Maths, "The Cicada's 17-Year Life Cycle", http://www.murderousmaths.co.uk/cicadas.htm

[Tab16] Esteban G. Tabak, "Why Do Cicadas Have Prime Life Spans?", http://www.cims.nyu.edu/~eve2/cicadas.pdf

Voyages Through Animalspace

[Daw05] Richard Dawkins, *The Ancestor's Tale*, Mariner Books, 2005

A Heap of Seagulls

[Daw05] Richard Dawkins, *The Ancestor's Tale*, Mariner Books, 2005

[Spe08] Jeff Speaks, "The Sorites Paradox", http://www3.nd.edu/~jspeaks/courses/2007-8/93914/_HANDOUTS/sorites.pdf, April 22 2008

[WSo16] Wikipedia article, "Sorites Paradox", http://en.wikipedia.org/wiki/Sorites_paradox

Chapter 12: Puzzling Paradoxes

You Can Cross the Road

[WAr16] Wikipedia article, "Archimedes", http://en.wikipedia.org/wiki/Archimedes#Mathematics

[WZe16] Wikipedia article, "Zeno's Paradox", http://en.wikipedia.org/wiki/Zeno%27s_paradox

Four Dimensional Greek Warships

[WSh16] Wikipedia article, "Ship of Theseus", http://en.wikipedia.org/wiki/Ship_of_theseus

[WVe16] Wikipedia article, "Velvet Underground", http://en.wikipedia.org/wiki/Velvet_underground

A Christmas Surprise

[Cho98] Timothy Y. Chow, "The Surprise Examination or Unexpected Hanging Paradox", http://www-math.mit.edu/~tchow/unexpected.pdf, 1998

[Gev04] Uri Geva, "The Surprise Paradox, Is It a Paradox?", MathVentures, http://www.mathventures.com/mathematicool/paradoxes/SurpriseParadox.htm, 2004

[WSu16] Wikipedia article, "Unexpected Hanging Paradox", http://en.wikipedia.org/wiki/Unexpected_hanging_paradox

Resolving the Grandfather Paradox

[Cha09] Mark Chacksfield, "Large Hadron Collider Sabotaged by Time Travel?", Tech Radar, http://www.techradar.com/news/world-of-tech/future-tech/large-hadron-collider-sabotaged-by-time-travel--642543, October 14 2009

[WGr16] Wikipedia article, "Grandfather Paradox", http://en.wikipedia.org/wiki/Grandfather_paradox

[WNo16] Wikipedia article, "Novikov Self-Consistency Principle", http://en.wikipedia.org/wiki/Novikov_self-consistency_principle

[WWo16] Wikipedia article, "World Line", http://en.wikipedia.org/wiki/World_line

A Million Dollar Choice

[Kie00] Frank Kiekeben, "Newcomb's Paradox", http://www.franzkiekeben.com/newcomb.html, 2000

[WNe16] Wikipedia article, "Newcomb's Paradox", http://en.wikipedia.org/wiki/Newcomb%27s_paradox

The Painter's Paradox

[WGa16] Wikipedia article, "Gabriel's Horn", http://en.wikipedia.org/wiki/Gabriel%27s_horn

The Monty Haul Paradox

[WMo16] Wikipedia article, "Monty Hall Paradox", http://en.wikipedia.org/wiki/Monty_Hall_Paradox

A Mathematical Nuclear Bomb

[WGo16] Wikipedia article, "Godel's Incompleteness Theorems", http://en.wikipedia.org/wiki/Gödel%27s_incompleteness_theorems

[WPr16] Wikipedia article, "Principia Mathematica", http://en.wikipedia.org/wiki/Principia_Mathematica

Chapter 13: Rethinking Reality

Answering All Possible Questions

[Bar16] University of Barcelona, "Who Was Ramon Llull?", http://quisestlullus.narpan.net/eng/611_info_eng.html#

[Bon16] Anthony Bonner, "What Was Llull Up To?", http://www.ramonllull.net/sw_studies/studies_original/compbon.html

[Cro05] John N. Crossley, "Raymond Llull's Contributions to Computer Science", http://www.csse.monash.edu.au/publications/2005/tr-2005-182-full.pdf, 2005

[WRa16] Wikipedia article, "Ramon Llull", http://en.wikipedia.org/wiki/Ramon_Llull

Sacrificing a Goat to Calculus

[Jon89] David E. H. Jones, "The World of Mathematics (review)", http://www.project2061.org/publications/rsl/online/TRADEBKS/REVS/WORLMATH.HTM, 1989

[WAn16] Wikipedia article, "The Analyst", http://en.wikipedia.org/wiki/The_Analyst

[WGe16] Wikipedia article, "George Berkeley", https://en.wikipedia.org/wiki/George_Berkeley

[WLi16] Wikipedia article, "Limits", http://en.wikipedia.org/wiki/Limit_of_a_function

Grue and Bleen

[WGr16] Wikipedia article, "Grue and Bleen", http://en.wikipedia.org/wiki/Grue_and_Bleen

Hippos in My Basement

[Hod16] Mark Hodes, "Thinking Well and Thinking Logically", http://www.skepticfiles.org/skep2/thinking.htm

Is There a Hippopotamus In This Podcast?

[Mac93] J. F. MacDonald, "Russell, Wittgenstein and the problem of the rhinoceros", Southern Journal of Philosophy 31 (4), 1993

[Mar10] Andy Martin, "Beyond Understanding", New York Times, http://www.andyross.net/autism.htm, November 21 2010

[Rys07] Simon van Rysewyk, "Wittgenstein: If a lion could speak, we could not understand him", https://wittgensteinforum.wordpress.com/2007/06/13/wittgenstein-if-a-lion-could-speak-we-could-not-understand-him-pi-p223/, June 13 2007

[WWi16] Wikipedia article, "Wittgenstein", http://en.wikipedia.org/wiki/Wittgenstein

I Have Lied to You

[Bur09] Ben Burgis, "A Half-Baked Thought About The Lottery Paradox, the Preface Paradox and the Philosophy of Science", http://blogandnot-blog.blogspot.com/2009/06/half-baked-thought-about-lottery.html, June 17 2009

[WLo16] Wikipedia article, "Lottery Paradox", http://en.wikipedia.org/wiki/Lottery_paradox

[WPr16] Wikipedia article, "Preface Paradox", http://en.wikipedia.org/wiki/Preface_paradox

[WRo16] Wikipedia article, "Rounding Errors", http://en.wikipedia.org/wiki/Rounding_error

Does This Podcast Exist?

[Bos03] Nick Bostrom, "Are You Living in a Computer Simulation?", Philosophical Quarterly Vol. 53, No. 211, 2003

[Gay12] Damien Gayle, "Do We Live in the Matrix?", Daily Mail, http://www.dailymail.co.uk/sciencetech/article-2216189/Do-live-Matrix-researchers-say-way-prove-do.html, October 11 2012

[Gra11] Ed Grablanowski, "You're Living in a Computer Simulation, and Math Proves It", IO9, http://io9.gizmodo.com/5799396/youre-living-in-a-computer-simulation-and-math-proves-it, May 7 2011

A Universe Made of Math

[Fra08] Adam Frank, "Is The Universe Actually Made of Math?", Discover Magazine, http://discovermagazine.com/2008/jul/16-is-the-universe-actually-made-of-math, June 16 2008

[Hos08] Sabine Hossenfelder, "Discover Interview With Tegmark", Back Reaction, http://backreaction.blogspot.com/2008/06/discover-interview-with-tegmark.html, June 17 2008

Mathematical Immortality

[WMa16] Wikipedia article, "Many-Worlds Interpretation", http://en.wikipedia.org/wiki/Many-worlds_interpretation

[WSi16] Wikipedia article, "The Singularity", http://en.wikipedia.org/wiki/The_singularity

Index

© Erik Seligman 2016

E. Seligman, *Math Mutation Classics*, DOI 10.1007/978-1-4842-1892-1

Get the eBook for only $5!

Why limit yourself?

Now you can take the weightless companion with you wherever you go and access your content on your PC, phone, tablet, or reader.

Since you've purchased this print book, we're happy to offer you the eBook in all 3 formats for just $5.

Convenient and fully searchable, the PDF version enables you to easily find and copy code—or perform examples by quickly toggling between instructions and applications. The MOBI format is ideal for your Kindle, while the ePUB can be utilized on a variety of mobile devices.

To learn more, go to www.apress.com/companion or contact support@apress.com.